P9-ARV-682

T.R.R.G.
$9.95

LARE

The Rockets' Red Glare

An illustrated history of rocketry through the ages

**Wernher von Braun
and Frederick I. Ordway III**

At the turn of the nineteenth century, the rocket had just begun to be an important element in warfare, its fiery red flare inspiring the American anthem during the War of 1812. Rockets were fired at Waterloo in 1815, in China during the Opium Wars of the 1840s, in the Crimea in the 1850s, and in Africa during the latter half of the 1800s. By the end of the nineteenth century the rocket was a familiar sight from one end of the earth to the other.

By the first decades of our own century, a fresh breed of inventors was dreaming of a new use for rockets—manned travel beyond the atmosphere. Their attempts led to one of the most far-reaching and exciting accomplishments in human history. Today we routinely watch space launches on our television sets. Tomorrow anyone who is healthy enough to fly in an airplane will be able to

THE ROCKETS' RED GLARE

◎

Wernher von Braun
and
Frederick I. Ordway III

board NASA's latest project, the Space Shuttle, and ride personally into orbit.

Two of the world's foremost authorities on rocketry trace its development from antiquity to modern times in this beautifully illustrated book. They describe the varied uses of rockets for fireworks and signals, harpooning whales, carrying mail, lifesaving, and high altitude photography. They also reveal little-known details about the invention of different types of rockets, including a personal account of Wernher von Braun's work on the V-2 during World War II. Rare lithographs, drawings, and color photos further enhance this volume.

LEDAY
YORK

Library of Congress Cataloging in Publication Data

Von Braun, Wernher, 1912–
The rockets' red glare.

Bibliography: p. 191.
Includes index.
1. Rocketry—History. I. Ordway, Frederick Ira, 1927– joint author. II. Title.
TL781.V59 621.43'56'09
ISBN 0-385-07847-1
Library of Congress Catalog Card Number 75–6162

Printed in the United States of America
First Edition

ACKNOWLEDGMENTS

The authors are most grateful for the assistance of the following organizations in making this book possible:
Atomic Energy Commission, Germantown, Maryland
Bibliothèque Nationale, Paris
Bibliothèque Royale, Brussels
Boeing Aerospace Company, Seattle, Washington
British Museum, London
Chrysler Corporation, Detroit, Michigan, and Huntsville, Alabama
City Museum, Helsinki
Department of the Air Force, Washington, D.C.
Department of the Army, Washington, D.C.
Department of Defense, Washington, D.C.
Department of the Navy, Washington, D.C.
Deutsches Museum, Munich
General Dynamics Corporation, Fort Worth, Texas, and Huntsville, Alabama
Hughes Aircraft Company, Aerospace Group, Culver City, California
Imperial War Museum, London
Kungliga Armé Museum, Stockholm
Lockheed Missiles and Space Company, Sunnyvale, California
McDonnell-Douglas Corporation, Huntington Beach, California, and Huntsville, Alabama
Ernst McMeans Studios, Huntsville, Alabama
Martin Marietta Corporation, Baltimore, Maryland; Denver, Colorado; and Huntsville, Alabama
Metropolitan Museum of Art, New York, New York
Musée de l'Air, Paris
Museo Nazionale d'Artiglieria, Turin
National Aeronautics and Space Administration, Washington, D.C., Cape Canaveral, Florida, and Huntsville, Alabama

National Archives, Washington, D.C.
National Maritime Museum, Greenwich, England
Northrop Corporation, Los Angeles, California
Numax Electronics, Inc., Hauppauge, New York
Philco-Ford Corporation, Newport Beach, California
Raytheon Company, Bedford, Massachusetts
Rockwell International, El Segundo, California
Rotunda Museum, Royal Artillery Establishment, Greenwich, England
Smithsonian Institution, National Air and Space Museum, Washington, D.C.
Stonington Historical Society, Stonington, Connecticut
Tøjhusmuseet, Copenhagen
United Technology Center, Sunnyvale, California
U. S. Army Missile Command, Huntsville, Alabama
Western Electric Company, New York, New York
West Point Museum, West Point, New York
Whaling Museum, New Bedford, Massachusetts

Especial gratitude is expressed to Maria Cooper Janis and to Harry H.-K. Lange for preparing original illustrations and to Richard B. Hoover for invaluable photographic assistance. The authors are also indebted to William H. Bond, Thomas L. Burkett, Elie Carafoli, Frederick C. Durant, Morton T. Eldridge, Rolf Engel, David G. Harris, Warner S. Ray, Mitchell R. Sharpe, C. Holley Taylor-Martlew, A. Ingemar Skoog, Miecryslaw Subotowicz, James R. Ware, and Frank H. Winter.

CONTENTS

THE ROCKETS' RED GLARE

The Rocket Man, by Cornelis Dusart. Dutch, seventeenth century. (Courtesy National Maritime Museum)

1

THE WESTWARD SPREAD OF THE ROCKET

The history of rocketry in the West is interwoven with the history of gunpowder, artillery, and pyrotechnics—gunpowder because it has long been the principal ingredient of rocket devices, artillery because cannon shells and rocket projectiles perform similar functions for the military, and pyrotechnics because rockets have traditionally been a staple in the art of employing fire for recreational, utilitarian, and military purposes.

Incendiary devices, though not rockets, have been used for more than two thousand years and perhaps up to three thousand. Soldiers threw boiling pitch and other burning substances against their attackers in Assyrian times, and the Greek army used incendiary missiles such as fire pots and fire arrows from the fourth century B.C. onwards. Tacitus mentioned fire lances while Vegetius, writing some seven hundred years later, reported an incendiary mixture containing sulfur, resin, bitumen, and tow (a fiber of flax, hemp, or jute) soaked in petroleum. In A.D. 399, Claudianus of Alexandria refers to what may have been a public fireworks display: "Twisting and turning globes of fire" that "ran about in different directions over the planks without burning or even charring them." Some historians dispute the firework interpretation, suspecting that Claudianus was describing an optical effect caused by a series of mirrors.

During the Arab siege of Constantinople at the end of the seventh century, the besieged Greek soldiers introduced the terrible "Greek fire" to the eastern Mediterranean world. Edward Gibbon writes, "the Saracens were dismayed by the strange and prodigious effects of artificial fire" on their forces when they attacked the city. The Byzantine Emperor Leo III the Isaurian (680–741) related that Greek fire was ejectable and that it was thrown out by "siphons" toward enemy naval targets. "We have divers ways of destroying the enemy's ships," he stated, "as by means of fire prepared in tubes, from which they issue with a sound of thunder, as with a fiery smoke that burns the vessels on which they are hurled." Warriors used "hand siphons," he added, to "throw the prepared fire into the faces of the enemy." The British professor of chemistry and historian of Greek fire and gunpowder J. R. Partington reports that a hand pump for projecting Greek fire is illustrated in an eleventh-century Vatican manuscript.

However it may have been projected, Greek fire helped the Byzantines repulse the attempt of Igor the Russian to capture Constantinople in 941, and later, in 1103, the Greeks put the Pisans to rout near the island of Rhodes by expertly employing the same weapon. Anna Comnena, daughter of Emperor Alexius I Comnenus, described the action and what she claimed was the composition of the incendiary agent.

> This fire [she wrote] they made by the following arts. From the pine and certain such evergreen trees inflammable resin is collected. This is rubbed with sulfur and put into tubes of reed, and is blown by men using it with violent and continuous breath. Then in this manner it meets the fire on the trip and catches light and falls like a fiery whirlwind on the faces of the enemy.

Since the Byzantines regarded the precise composition of Greek fire as a vital military secret, Anna Comnena did not mention that petroleum was an ingredient. By the time of the Crusades, however, knowledge of Greek fire was widespread and attackers and defenders alike hurled it in battle. During the siege of Al-Mansūra in the Nile Delta in 1249—at which time Sultan al-Mu'azzam's forces captured the French King Louis IX —the Arabs fired Greek fire with ballistas. A thunderous noise was produced and "such a light that we could see in our camp as clearly as in broad day" (Jean, Sire de Joinville, in *Histoire de Saint Loys*).

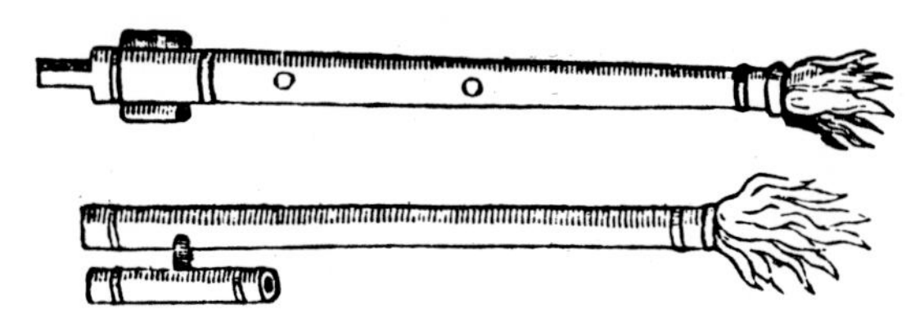

Rockets described by Marcus Graecus in his thirteenth-century manuscript *Liber Ignium.*

A jellylike material, Greek fire must have contained distilled petroleum along with sulfur, resin, and pitch "mixed with certain less important and more obscure substances," according to the British historian C. W. C. Oman. Though it and other pyrotechnic mixtures were common in the eastern Mediterranean, the reactive forces of incendiaries were probably not applied to the propulsion of projectiles prior to the thirteenth century. A new chapter in the history of pyrotechnics had first to be written.

This new chapter was a small manuscript entitled *Liber Ignium ad Comburendos Hostes,* or *Book of Fire.* Traditionally it has been attributed to a writer named Marcus Graecus, but scholars are divided as to whether or not he was the author—or even if he existed at all. Whoever wrote or compiled the work prepared thirty-five Greek pyrotechnic recipes dating from the mid-eighth to the end of the thirteenth centuries. (The manuscript is a collection of secret and special recipes that was eventually brought together in a compendium, possibly by a Jew or by a Spaniard. No evidence exists that Marcus Graecus was a Greek or that the original manuscript was written in that language.)

Liber Ignium contains references to flying devices and to a vital ingredient of gunpowder, saltpeter:

> Note there are two compositions of fire flying in the air. This is the first. Take one part of colophonium [a resin from Colophon in Greece] and as much native sulfur, six parts of saltpeter. After finely powdering, dissolve in linseed oil or in laurel oil, which is better. Then put into a reed or hollow wood and light it. It flies away suddenly to whatever place you wish and burns up everything.

Military historian Colonel Henry W. L. Hime called this description of a rocket "as definite and precise as many a recipe of the seventeenth century."

Marcus Graecus offers a "second kind of flying fire" and spells out how to make it:

> Take one pound of native sulfur, two pounds of linden or willow charcoal, six pounds of saltpeter, which three things are very finely powdered on a marble slab. Then put as much powder as desired into a case to make flying fire or thunder.

If one wanted flying fire, he cautioned, one should make the case long and narrow and pack the powder firmly. Still a third recipe states that for one part sulfur there should be three parts charcoal, and for one part charcoal, three parts saltpeter.

The introduction of saltpeter into pyrotechnic mixtures presaged the eventual demise of Greek fire, though the latter persisted in Europe well into the fourteenth century. When added to sulfur and charcoal, saltpeter (which was also known as niter and chemically as potassium nitrate) forms an explosive mixture capable of hurling artillery shells and of propelling rockets over ranges far beyond that attainable by Greek fire. It is impossible to determine how Marcus Graecus came to learn about saltpeter; his several recipes listing it as an ingredient date from the latter part of the thirteenth century. The Spanish Arab al-Batyār mentions saltpeter, calling it "snow of China" and suggesting it was introduced into Europe via the Middle East. Al-Baytār died in 1248.

Both Albertus Magnus, the German scientist, and Roger Bacon, the English one, fully realized in the thirteenth century the importance of saltpeter and recorded recipes for explosive powders. In Partington's opinion, these two men of the Middle Ages "had intellects not inferior to Galileo's and Newton's, and had they lived in the time of Galileo and Newton they would probably have been outstanding scientists." Albertus Magnus was born between 1193 and 1207 in or near Lauingen in Swabia, and in 1229 became a Dominican priest. Rising to become Bishop of Ratisbon from 1260 to 1262, he soon thereafter retired to writing and teaching throughout much of Germany. Of his incredible thirty-eight-volume literary output, *De Mirabilis Mundi* interests us most, for in it he recorded recipes for gunpowder similar or identical to those appearing in *Liber Ignium.* In fact, they may have been taken from it; for example, the recipe for flying fire is almost a verbatim copy of the fire book's recipe for a "second kind of flying fire." Writes Albertus Magnus:

> Take one pound of sulfur, two pounds of willow charcoal and six pounds of saltpeter, which three things grind finely on a marble stone. Then put as much as you wish into a paper case to make flying fire or thunder. The case for flying fire should be long and thin and well filled with powder, that for making thunder short and thick and half filled with powder.

Since the author spent most of his retirement in a cloister in Cologne, he could easily have had access to a late—though not the latest—version of *Liber Ignium* and its gunpowder recipes.

Born a few years after Albertus Magnus, Roger Bacon started life in Ilchester, Somerset, in the year 1214. Though closely associated with the academic world at Oxford University and a member of the Franciscan order, many of his literary works were not published until long after his death—the *Opus Majus* in 1733 and *Opus Tertium* and part of *Opus Minus* in 1859! More than anyone before him, Bacon advocated the scientific method, and insisted that his myriad recipes, some almost magical in nature, be tested before being accepted. Like his contemporaries, Bacon treated as science many aspects of alchemy, astrology and necromancy.

Between 1257 and 1265, Bacon composed *De Secretis Operibus Artis et Naturae et de Nullitate Magiae,* eleven letters (*epistola*) on the marvelous power of art and of nature and "the nullity of magic." They predate the portions of *Liber Ignium* dealing with gunpowder, though not other parts of that manuscript nor Albertus Magnus' *De Mirabilibus Mundi.*

In describing gunpowder, Bacon was less than lucid. At the end of the sixth letter, for example, we find "Annis Arabum 630 . . . Item pondus totum sit 30. Sed tamen sal petrae LURU VOPO VIR CAN VTRIET sulphuris; et sic facies tonitruum et coruscationem, si scias artificium." This has subsequently appeared in English as (1): "Notwithstanding, thou shalt take salt-petre, *Loro vopo vir can vtri,* and of sulfur, and by this means make it both to thunder and to lighten"; and as (2): "But get of saltpetre LURU. VOPO *Vir Can Utriet* sulfuris and so you may make Thunder and Lightning if you understand the artifice." In working out what he supposed to be an anagram, Colonel Hime rearranged the mysterious letters and combined them into groups which yielded a recipe for gunpowder comprising seven parts saltpeter, five of young hazelwood charcoal, and five of sulfur. This mixture, Hime believed, would explode "if you know the trick"—that is, if you were certain that the saltpeter was pure, that all the ingredients were well mixed, that the resulting powder was kept dry, and that is was not overly packed. According to Hime, Bacon discovered gunpowder while he was experimenting with incendiary compositions prepared with pure instead of impure saltpeter. Hime speculates that "the mixture exploded unexpectedly and shattered all the chemical apparatus near it . . ."

Gunpowder is described elsewhere by Bacon. For example, he writes, "From the force of the salt called saltpetre so horrible a sound is produced by the bursting of so small a thing, viz. a small piece of parchment, that we perceive it exceeds the roar of strong thunder and the flash exceeds the greatest brilliancy of the lightning" (*Opus Majus*). Further, "By the flash and combustion of fire, and by the horror of sounds, wonders can be wrought, and at any distance that we wish—so that a man can hardly protect himself or endure it" (*Opus Tertium*). Bacon, then, knew how to make gunpowder and what its effects were. He also experimented with other explosive mixtures, refining his understanding as he progressed. However, we do not know whether he appreciated the propulsive properties of gunpowder, or whether he built and flew simple powder rockets. Furthermore, we cannot substantiate Hime's view that Bacon *discovered* gunpowder, or if—as is more likely—he learned of it through recipes filtering into Medieval Europe from the East. Bacon himself never claimed the discovery.

Whatever the true facts may be, Bacon's research soon led to the widespread use of gunpowder in peaceful and military enterprises. Traditionally, the name "Black Berthold," or "Berthold Schwartz," or "Bertholdus Niger," has turned up as (1) the in-

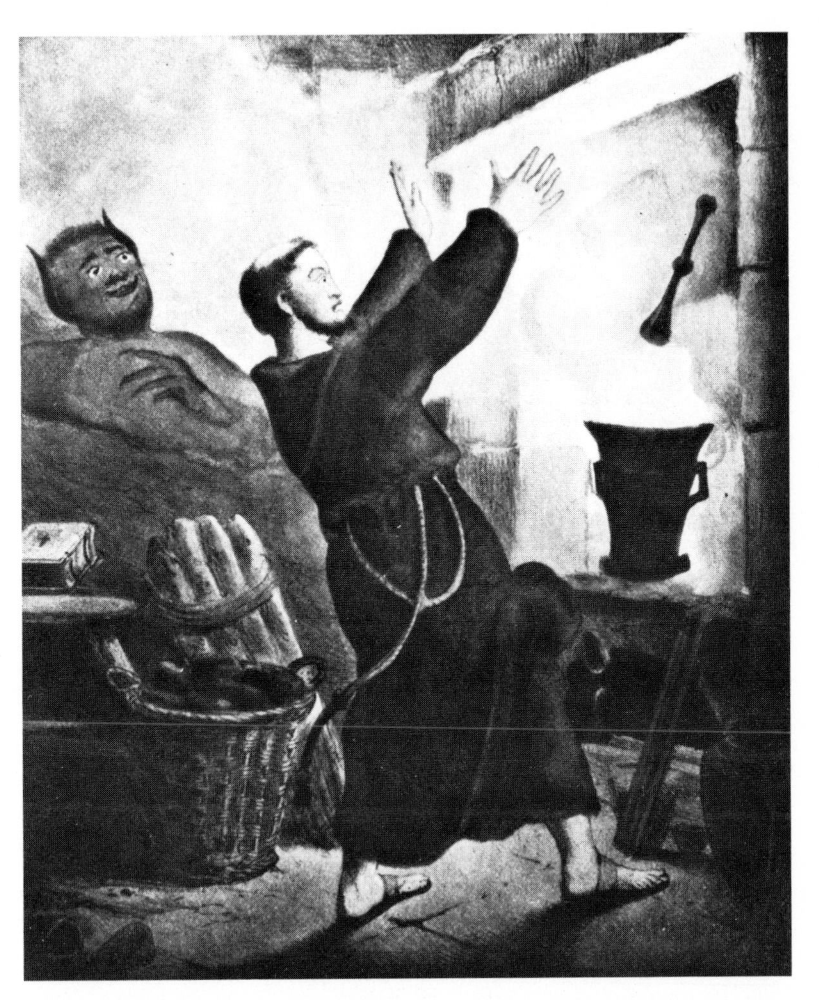

The mythical Black Berthold (Berthold Schwartz, Bertholdus Niger, or Niger Bertholdus) was believed by some chroniclers to have invented gunpowder while seeking to discover a gold paint. In the process, according to Oscar Guttmann, "the crucible or mortar was blown into the air." The name "Black" came about as gunpowder-making was considered a "black art." Here we see Berthold, the mortar, and the figure of the devil.

A page from Konrad Kyeser von Eichstädt's *Bellifortis* (1395–1405), showing a pyrotechnician, a crude launcher, and, at upper right, a flying rocket. The Oriental garment suggests that Kyeser and his artist leaned heavily on Arabian-Byzantine sources for inspiration. The peculiar way that the guide stick is hanging is evidence that the artist was not too familiar with his subject. The text, however, is full of interesting detail. Kyeser explains that "an axial channel must be drilled into the rocket's grain, or driving charge, in order to provide enough combustion surface, exhaust gas, and thrust for adequate acceleration." He also states that "a rocket must carry a separate incendiary payload, as the residue of the propulsive charge is not very effective to set a target on fire." (This is a clear indication that Kyeser was thinking of military applications.) "If, instead of an incendiary effect," he added, "a loud bang is desired, the payload should consist of a little bomb ignited by the rocket charge at burn-out." (Courtesy Deutsches Museum)

ventor of gunpowder, and/or (2) the inventor of the cannon. Supposedly a German, though sometimes referred to as a Dane or even a Greek, he may have lived in the mid-fourteenth century. More likely, Black Berthold was a legendary figure invented, according to Partington, "solely for the purpose of providing a German origin for gunpowder and cannon."

As solid as Berthold was ephemeral, Konrad Kyeser von Eichstädt left an important military manuscript composed between 1395 and 1405 (the year in which he probably died). Entitled *Bellifortis*, it runs 243 pages plus 140 parchment leaves containing illustrations of all manner of military devices. One of these is a rocket launcher, another a rocket flying through the air. Although there is no directly emplaced stick to help guide the rocket's trajectory, the text insists that one should be provided. Moreover, Kyeser explained that the rocket was propelled by hot exhaust gases and that the propellant charge was placed in the bore of the rocket. He also noted that the walls of the container in which the powder was packed had to be impervious to the escaping gases.

Though important to the history of fireworks, the fifteenth-century *Feuerwerkbuch* shed little light on rockets per se. Surviving in manuscript form in Florence, Vienna, Dresden, Berlin, and Paris, it includes later interpretations that make it difficult to date events precisely. The treatise describes saltpeter, sulfur, charcoal, gunpowder, and "special" powders in detail.

If the fireworks book was vague about the use of gunpowder in rockets, Giovanni da Fontana (who probably lived between 1395 and 1455) was not. In a beautifully illustrated manuscript entitled *Bellicorum Instrumentorum Liber cum Figuris et Fictivis Literis Conscriptus*, Fontana, a Venetian engineer, writes of flying rockets, jet-propelled carts, rockets that skip along and even go under water, and other delights. In another work, *Metrologum da Pisce, Cane et Volcure*, he describes the quantities of powder necessary to fire different types of rockets, as well as ways of making rocket-powered dragons that would belch forth fire and obnoxious gases.

During the fifteenth, sixteenth, and seventeenth centuries, manuscripts and books appeared with increasing frequency that were —and are—of importance in the history of rocketry. Leonardo da

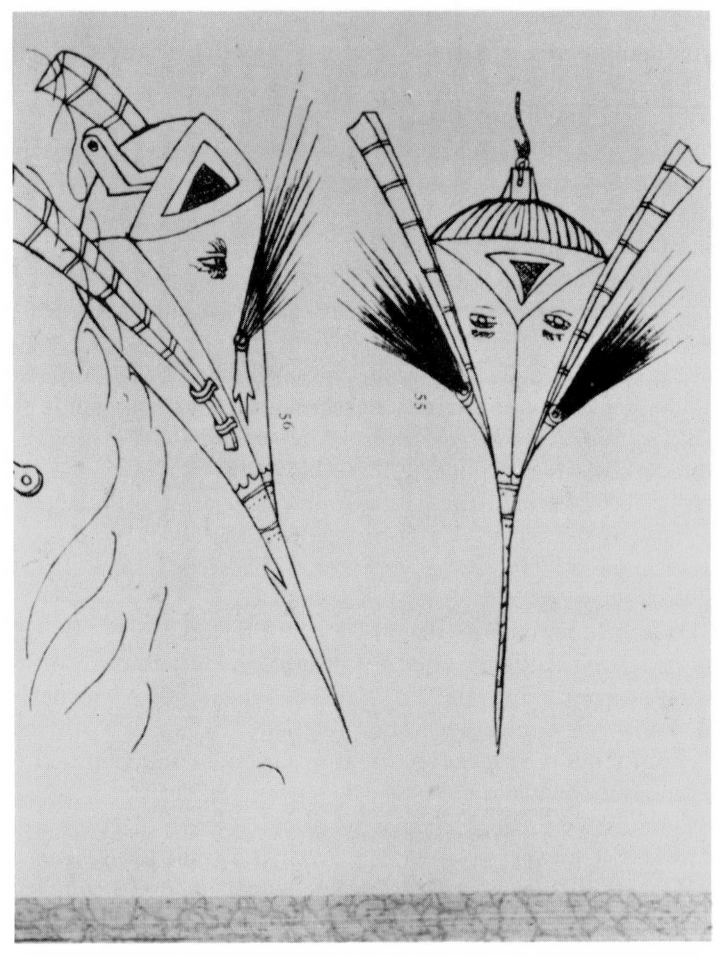

Giovanni da Fontana, fifteenth-century Venetian engineer and author of the *Bellicorum Instrumentorum Liber* and other works, described rocket-powered torpedoes that could skip along and even go under the surface of the water en route to their targets. (Courtesy Maria Cooper Janis)

Vinci (1452–1519) prepared drawings of rockets, which appear in the *Codice Atlantico* in Milan. In the great *De la Pirotechnia*, the Sienese Vanoccio Biringuccio (1480–1539) describes the making of both gunpowder and fireworks, while another Italian, Niccolo Tartaglia (born in Brescia c. 1500), provided recipes for constructing civilian and military fireworks of many types. Across the Alps, Leonhart Fronsperger of Frankfurt-am-Main composed a work on artillery which appeared in 1557 and in which rockets were said "to go high into the air, to give off beautiful fire," and then to use themselves up "without any harm." About fifteen years later, Samuel Zimmermann of Augsburg wrote on all manner of fireworks, followed by Johann Schmidlap who described nonmilitary pyrotechnics (1591). A Spanish artillery captain, Diego Ufano, took up the other side of the matter and concentrated on military rockets in the early 1600s.

About a decade ago, an interesting manuscript came to light in the town of Sibiu in the central part of Romania. Written in sixteenth-century German by Conrad Haas, it has been carefully examined by Doru Todericiu of the University of Bucharest and by Elie Carafoli of the Academy of the Socialist Republic of Romania. They report that Haas served as chief of the Arsenal of Artillery in Sibiu from 1529 to 1569 and that the calligraphy of the manuscript is very accurate. According to Carafoli, "Even the preliminary pages record the interest of the Sibian author in the construction of multistage rockets. It is with a surprising clearness that Conrad Haas attained, even in 1529 (at the beginning of his activity in the field of practical pyrotechnics), to the conception of the multi-staged rocket." The manuscript shows his "double" or two-stage rocket and his three-stage "variant." To the right of the three-stage rocket is another design featuring two chambers exhausting in the opposite directions. One rocket was to fire upwards, the other igniting later to bring the missile back "as close as possible to the taking-off point." Haas also developed a swept-back guidance fin. In a moment of fancy, he sketched a small, one-story house attached to a rocket. Carafoli speculates that it "might be interpreted as a naïve prefiguration of the idea of a future manned flight."

To the north, in Poland, T. Przypkowski suspects the monk named Seweryn, who lived c. 1380, had in mind rockets when he wrote of applying powder to propel "tubes." According to research undertaken by Mieczylaw Subotowicz of the Institute of Physics, M. Curie-Skłodowska University, in Lublin, the first fully authenticated work in the Polish language in which rockets

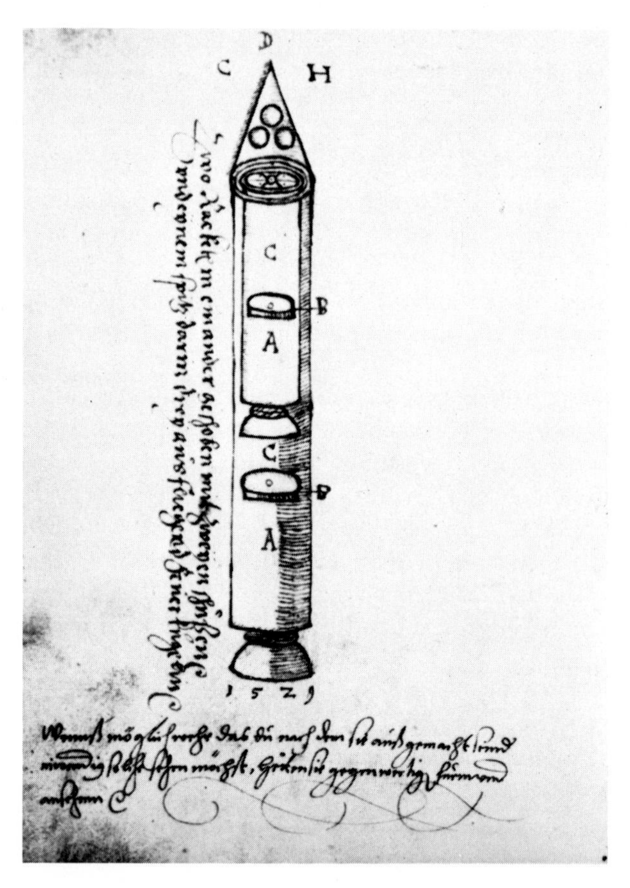

A sixteenth-century German two-stage rocket, sketched by Conrad Haas, pyrotechnic expert. The powder charge of the lower stage is first ignited, and the rocket takes off. Upon burn-out of the first stage, the flame is conveyed to the propellant in the second stage, which commences firing. According to Carafoli, "the separation of the first stage after the extinction of the respective propellant appears to be no longer necessary. The conception regarding the manner in which this type of rocket is running includes the quite ingenious idea of the integral consumption of the first-stage engine while the propellant is being burned up. To this end, Conrad made the rocket cover out of paper impregnated with various substances, which burns up in proportion as the propellant is consumed, such that, when the first stage is exhausted, the second engine remains on the trajectory as an independent rocket." A page from the sixteenth-century manuscript is reproduced above. (Courtesy E. Carafoli)

are described is M. Bielski's *Sprawa Rycerska,* published in Krakow in 1569. Sometime around 1600 Walenty Sebisch (1577–1657) sketched a number of rocket devices, including one with delta-shaped fins. A little later (between 1630 and 1635) a Venetian citizen, A. Dell'Aqua, prepared a manuscript entitled *Praxis Reçzna Dzicła* while in the service of the Polish government. Although it dealt primarily with the "hand production of guns," the manuscript also described a number of rockets including a two-stage design whose upper stage contained, in turn, five small rockets.

Despite the interest of military experts, there is no evidence of widespread use of war rockets in Renaissance Europe. The first engagement in which such rockets figured with any prominence was recorded by the early eighteenth-century historian Lodovico Antonio Muratori in *Rerum Italicarum Scriptores,* in which he wrote regarding a fourteenth-century battle that a defending tower that had previously resisted attack finally succumbed when set afire by a rocket. This event occurred at Chioggia, a seaport on an island at the southern end of the Venetian lagoon, during the 1379–80 war between Venice and Genoa. The word *rocchetta* came into use at about that time; during the Middle Ages in Italy, soldiers used a small, round piece of wood known as a *rocca* (bobbin, diminutive of distaff) to cover the sharp points of lances during mock combat. Since the shape of the rocca was similar to that of the new fire weapon, the word *rocchetta* gradually evolved, leading in turn to the later term *rocheta,* the English *rocket,* Russian *rakety,* and the German *rakete.* (Oddly, in modern Italian, rockets are known as *razzi.*)

Occasional use of rockets in other battles has been recorded—the forces of Jeanne d'Arc used war rockets in 1428 at Orleans, and in 1449 they figured in a battle against the English at Pont-Audemer—but they were more often studied as a curiosity by military engineers than fired in the field. In several books appearing between 1598 and 1630, Jean Appier of Lorraine (who used the pseudonym Hanzelet) looked into an astonishing variety of pyrotechnic devices, ranging from military rockets fitted with iron heads, to rockets armed with grenades to terrify cavalry, to rockets that ejected noise-making firecrackers. His eight-pound

Simple wooden launcher used during the sixteenth and seventeenth centuries, fitted with a "rest" made of "white tinne plate." Here a rocket is being prepared for launch against a fortified castle. (From Hanzelet's *La Pyrotechnie Militaire,* 1598)

iron-headed rockets contained eight pounds of saltpeter, 2¾ pounds of charcoal, and 1¼ pounds of sulfur.

Popular with Appier and other seventeenth-century pyrotechnic experts were rockets used to propel objects strung out along cords between two or more structures. Typically, such structures were battlements in scenic or display representations of castles. An excellent example of such a device appears in *Récréations Mathématiques,* a book published in Rouen in 1624 and 1630. Compiled from Greek and Latin sources either by Henry van Etten or his teacher Jean Leurechon, it was translated into English by William Oughtred and published in London in 1633. Dragons were the most popular figures for a "rocket-on-a-cord" display (see illustration of similar arrangement on page 68):

> Take small *Rockets,* and place the tayle of one to the head of the other, upon a *Cord* according to your fancie, as admit the *Cord* to be *ABCDEFG.* give fire to the Rocket at *A,* which will fly to *B,* which will come backe againe to *A,* and fire another at *C,* that will fly at *D,* which will fire another there, and fly to *E,* and that to *F:* and so from *F* to *G;* and at *G,* may be placed a pot of fire, viz. *GH:* which fired will make good sport, because the *Serpents* which are in it will variously intermix themselves in the *Ayre,* and upon the ground, and every one will extinguish with a report: and here may you note that upon the *Rockets* may be placed fierie *Dragons Combatants,* or such like to meete one another, having lights placed in the Concavity of their bodies, which will give great grace to the action.

Athanasius Kircher, Lieutenant-General of Ordnance Casimir Siemienowicz of Poland, and Joseph Furttenbach wrote other important works. Siemienowicz's great *Artis Magnae Artilleriae* came out in Amsterdam in 1650 and was later expanded and translated into German and French. The English edition, which appeared in 1729 under the title *The Great Art of Artillery,* contains twenty-three excellent copperplate engravings. One of the rockets, adapted from a design originally prepared by Schmidlap, called for three stages or steps:

> . . . take *Rocket* B [the Polish general instructed] with the

Rocket with conical cap and attached guide stick popular in the sixteenth and seventeenth centuries. (Courtesy Harry H.-K. Lange)

> first *Rocket* A in it, and putting it in the Hollow of this Third, glue or paste them neatly together, and cover them all three with the Conic Head F, made either of Wood or Paper. You have the whole Order of this *Rocket* in the same *Figure,* distinguished by the Letter E.

Siemienowicz later explained that the first two rockets "will by the Third be carried up into the Air, where they perform their Parts; flying from one side to the other in oblique Directions; for they cannot ascend perpendicularly, for want of Sticks or a Counterpoise . . ."

He does add guiding sticks to some of his rockets in order to "assist them in the right Ascent." How long should a given stick be? The answer is precise:

> Add one to the Number of *Fingers* that constitute the Length of your *Rocket,* and multiply the product by the Length of the *Rocket,* and you will have the due Length of its *Stick:* for example; if the *Rocket* is 8 *Fingers* in Length, add 1 to them,

> and you will have 9; which Number multiplied by 8, which is the Length or Height of the *Rocket,* will give 72. You shall then tye a *Stick* of so many *Fingers* in Length to your *Rocket.*

Among many designs illustrated are skyrockets "*that mount up without* Sticks," which are affixed with four small wings "after the manner of the Feathers of an Arrow." One drawing has a surprisingly modern, delta-wing look. As for Furttenbach, he proposes in *Architectura Navalis* rockets armed with iron heads and lead balls for attacking pirates.

Many other illustrated works on pyrotechnics appeared during the seventeenth century, chief among them *The Mysteries of Nature and Art* by John Bate, *Pyrotechnia, or, A Discourse of Artificiall Fire-works* by John Babington, and Francis Malthus' *Treatise of Artificiall Fireworks Both for Warres and Recreation.* Malthus, sometimes known by the name François de Malthe, worked and wrote in France, serving as Commissaire Général des Feux et Artifice de l'Artillerie de France and as Capitaine Général des Sappes & Mines d'Icelle & Ingénieures, Armées du Roy. The *Treatise,* originally written in French, was translated into English by the author. It was followed by other works in the French language.

Meanwhile, in 1658, the English-language edition of Giovanni Battista della Porta's *Natural Magick* appeared in which we find a chapter on "Of Artificial Fires" tucked in between "Of Perfuming" and "Of Tempering Steel." The author describes a military rocket "of extraordinary largeness":

> ten foot long, and as wide as a mans head might go in: it was full of Fire-balls, Stones, and other matters, and put into a Gun, and bound to the lower part of the Cross-yard of a Ship, which was transported every way with cords, as the Souldiers would have it; and in Sea-fights was levelled against the Enemies Gallies, and destroyed them all almost . . .

Near the end of the seventeenth century, Robert Anderson published *The Making of Rockets.* A mathematician and silk-weaver from London and one of the early members of the Royal Society, he devoted much of his life to the improvement of the art of gunnery. Beginning in 1671, he conducted innumerable experiments at his own expense on Wimbledon Common and was highly satisfied with his work. "I am very well assured," he wrote. "I have done more, being a private person, than all the engineers and gunners with their yearly salaries and allowances, since the first invention of this warlike engine." Though he was referring to the cannon here, he was equally enthusiastic about his rockets, expressing himself in no uncertain terms to the Right Honourable Henry, Earl of Romney and Master General of His Majesty's Ordnance:

> I do not find the least Pretence of Thought of doing *that* which here is undertaken, viz. *To raise so great a heap of Fire, and to confirm the Fact by the greatest Proofs that can be had or wish'd for, which are Experiments and Demonstrations Mathematica* . . . I am emboldned to lay this mean Attempt of mine before your Lordship; trusting that it may at some time or other undergo a Trial, and by your Lordships Favour receive a meet Incouragement . . .

A century and a half later, another Englishman, Sir William Congreve, would propose in a similar vein a more advanced type of rocket ordnance. However, for the 150 years that followed Anderson, the rocket remained the plaything of royal fireworks makers.

2

FIREWORKS: THE JOY OF ROYALTY

From the fifteenth to the eighteenth centuries the rocket was associated with festive firework displays, which became progressively more elaborate, brilliant, and complex as the years went by. The Florentines and Sienese were the first Europeans to develop large-scale firework displays. They erected theaterlike structures decorated with flowers, emblems, statues, and paintings. From such structures, which might rise seventy or more feet in height, fireworks displays celebrated the Eve of Saint John, the Feast of the Assumption, and other special occasions. In Rome, the popes shot off fireworks in honor of Saint Peter and Saint Paul. Their favorite site, wrote the French artificer Amédée François Frezier was the Castel San Angelo, as the displays "could be seen from most of the city of Rome."

> It was towards the end of the sixteenth century that fireworks, introduced from Italy into France, brought to a close almost all the great demonstrations of joy with splendor and pomp. With all the solemnity of the occasion, fireworks highlighted birth and marriage ceremonies, the formal arrival of royalty, victories, and events in the careers of sovereigns and of nations.

So wrote Émile Magne in 1930 in his impressively illustrated book *Les Fêtes en Europe au XVII*^e *Siecle.* The splendid creations of pyrotechnicians enabled the French monarchy to take the pulse of the nation, familiarize itself with its subjects, provide a feeling of merrymaking and rejoicing among the masses, and achieve various political ends.

Fireworks celebrating the marriage of Louis Auguste, Dauphin of France, to Marie Antoinette of Austria in May 1770. Contemporary print.

Henri IV of France (1553–1610) started the tradition of setting off public fireworks with a flaming torch. This somewhat dangerous practice was carried on by Louis XIII (1601–43) and, for a while, by Louis XIV (1638–1715). The latter, however, soon took the sensible precaution of allowing one or another prince to perform this symbolic gesture.

For many years, the art of pyrotechnics was strictly controlled by royal engineers or engineers employed by the larger cities of France. So creative and inspiring were their displays that they could easily hold the attention of what otherwise might have been restless crowds. These pyrotechnic engineers or artificers were free agents insofar as the style of the creations was concerned; neither king nor moralists influenced their work, which was based on lively and dynamic presentations inspired more by events of the time than by the past. Their one concession to the past was mythology: artificers would often prepare the figures (from which their fireworks would be fired) of heroic size fashioned of plaster which covered a wooden frame that was later removed. Each figure would stand on a high platform that served

Spectacular night fireworks display in the court of the Bishop's Palace on the occasion of the 1603 trip of Henri IV of France to Metz.

wills. "From that time on," remarks Magne, "firework displays, formerly so diversified in their form, tend to take on the definite image of a monument, a temple, an arch of triumph. . . . No more romanticism, chimerical adventures, unexpected scenes." More and more, fireworks came to serve the political self-interests of kings. Ornate sides to the main structures were added, along with colonnades, peristyles, bas-reliefs—all suggesting the splendor and power of the throne.

Among the most impressive seventeenth-century fireworks displays were those celebrating the return of Louis XIV to Paris in 1660 and the birth of the dauphin in 1682 (with simultaneous festivities in Paris, Dijon, and Lyons). Louis XV (1710–74) kept the fireworks tradition alive in France by backing and encouraging the activities of the Ruggieri brothers from Italy and of Morel Torré. Alan St. H. Brock, British pyrotechnician, points out in his *Pyrotechnics* (1922) that the Ruggieris "were the first to rely chiefly on fireworks for the effect instead of using them merely to embellish a scenic or arthitectural structure." They conducted one of their greatest displays at Versailles in 1739 to celebrate the marriage of Madame la Première of France and Don Filipe of Spain. The capture of Château Grand in 1745 and Ypres in 1747 gave rise to other splendid displays before the Hôtel de Ville in Paris.

The city of Strasbourg gave a particularly impressive welcome to Louis XV on October 5, 1744. Infantry and cavalry were called up, their officers and men ornately dressed for the occasion and provided with splendid banners and flags. At four o'clock in the afternoon a great cry erupted as the king's carriage approached the Saverne entry gate. The royal entourage passed into the Saverne suburb and through a magnificent Corinthian-style arch of triumph flanked by Swiss guards. Along the sides of the arch stood a great statue of the king, with angels supporting the shields of the arms of France and of Navarre. Along graveled streets decorated with gorgeous tapestries the king made his way to the episcopal palace of Rohan where he subsequently lodged during the ten-day visit. According to a contemporary account, he was attended by twenty-four girls from fifteen to twenty years old "dressed in superb clothes cut according to the customs of the German residents of Strasbourg [and another] twenty-four girls dressed in accordance with French traditions." Gabriel Mourey (*Livre des Fêtes Françaises*) describes Louis' view from the balcony of his apartments as the event of the day—the spectacular fireworks display—unfolded.

> In front of the Rohan Palace, along the bank of the river Ill, a great forty- to fifty-foot-high, hundred-foot broad edifice had been constructed. It was an arch of triumph with seven arcades plus a larger central one that enclosed a statue showing the king arriving in Alsace on horseback. The words *Cum Domino Pax Ista Venit* were inscribed thereon, together with an inscription above it reading *Nec Pluribus Impar* and the rays of the sun.

A contemporary description has it that "the sun, the royal arms, various heraldic inscriptions, pyramids, and the fleurs-de-lys on the arch . . . suddenly [at nine o'clock at night] took flame, giving rise to a lively and brilliant display whose colors changed three times." Clouds of rockets flew into the air and other sorts of pyrotechnic devices went into action. Some were serpentine in shape and, after falling into the water and remaining there for some time, "suddenly came out, covered the surface of the river, and then dissipated into a thousand bursts." The fireworks lasted for three quarters of an hour, during which two orchestras played aboard illuminated and garlanded boats anchored in the river Ill. For days the feasts and celebrations continued, the public consuming vast quantities of roast beef, "all sorts of other meats and poultry, a profusion of bread, and copious quantities of wine." On the tenth day, Louis XV departed Strasbourg to the same ceremonies as those that had welcomed his arrival.

Although the Italians and French dominated the fireworks scene for several centuries, they had no claim for exclusivity in Europe. The Germans often shot off fireworks, one of their best-recorded displays occurring on July 8, 1667, at Pleissenburg at the behest of the Prince of Saxony. A couple of years later in Stockholm another impressive display took place on the occasion of Charles XI of Sweden's investiture with the British Order of the Garter. In England, fireworks became popular early in the seventeenth century. Shortly after the marriage in 1613 of James I's daughter to Prince Frederick the Elector Palatine, a long account of the fireworks event marking the occasion was prepared, entitled "The Manner of Fire-Workes Shew up upon the Thames." It read, in part:

> First, for a welcome to the beholders a peale of Ordnance like unto a terrible thunder ratled in the ayre. . . . Secondly, followed a number more of the same fashion, spredding so strangly with sparkling blazes, that the skie seemed to be filled with fire. . . . After this, in a most curious manner, an artificiall fire-worke with great wonder was seen flying in the ayre, like unto a fiery Dragon, against which another fierrie vision appeared flaming like to Saint George on Horsebacke, brought in by a burning Inchanter, between which was then fought a most strange battell continuing a quarter of an howre or more; the Dragon being vanquished, seemed to roar like thunder, and withall burst in pieces, and so vanished; but the champion, with his flaming horse, for a little time made a

Fireworks ship, probably to celebrate a Habsburg entry, by J. Neeffs. Notice the rockets being fired from the bowsprit. (The Metropolitan Museum of Art, Harris Brisbane Dick Fund, 1953)

shew of a tryumphant conquest, and so ceased.

After this was heard another ratling sound of Cannons . . . and forthwith appeared, out of a hill of earth made upon the water, a very strange fire, flaming upright like unto a blazing starre. After which flew forth a number of rockets so high in the ayre, that we could not chose but approve by all reasons that Arte hath exceeded Nature, so artificially were they performed. And still . . . the fire-workes danced in the ayre, to the great delight of his Highnes and the Princes . . . These were the noble delights of Princes, and prompt were the wits of men to contrive such princely pleasures.

The popularity of fireworks continued. The caption on one eighteenth-century print stated that during one evening a thousand skyrockets were used up, ranging from four to six pounds in weight, together with twenty-three rocket chests "each containing sixty rockets from one to four pounders." The cost of these plus such other items as light balls, shells, and Roman candles was £12,000.

The practice of celebrating special events and occasions with fireworks, widespread though it was, did not go unchallanged. Charles Lamb, for instance, in the nineteenth century, was dismayed at the damage done to Hyde and Green parks in London as a result of pyrotechnical displays and the jostling and milling about of the enthralled crowds. Yet he admitted in a letter he wrote to William Wordsworth on August 9, 1814 that

> After all the fireworks were splendent—the Rockets in cluster, in trees and all shapes, spreading about like young stars in the making floundering about in space . . . till some of Newton's calculations should fix them, but then they went out. Anyone who could see 'em and the still finer showers of gloomy rain fire that fell sulkily and angrily from 'em could go to bed without dreaming of the Last Day, must be as hardened an Atheist as ****

To a reporter for *The Times,* "the repetition of these things, with occasional pauses, for more than two hours became tedious to all." He was a distinct minority.

Despite a few grumblings, fireworks drew large crowds throughout most of the nineteenth century. C. T. Brock inaugurated a series of impressive displays at Crystal Palace in 1865, which were so successful that he soon built extensive manufacturing facilities at Nunhead in the London area. There, he began producing rockets and other pyrotechnic devices "on a scale never previously dreamt of in the trade." His brother Arthur took over the family business in 1881 and continued to hold firework displays at the palace until 1910.

Inevitably, European firework technology spread to the New World. Captain John Smith, governor of the New England colonies, records in his *The Generall Historie of Virginia, New-*

England that on the evening of July 24, 1608, ". . . we fired a few rockets, which flying in the ayre so terrified the poore Salvages [the Indians], that they supposed nothing unpossible we attempted; and desired to assist us." These firework rockets were brought from England, but beginning in the eighteenth century a native pyrotechnic industry took hold in the new country.

Scintillating, enthralling, and stimulating though fireworks were, they did not engender sustained development programs to improve the flight performance of rockets. Remarkably few changes occurred in manufacturing and launching practices until the nineteenth century. Preparations for war ultimately proved to be a greater incentive than preparations for festive celebrations.

Fireworks from Jean Thevenot's *Travels in the Levant, 1724,* by Jan Luyken. (The Metropolitan Museum of Art, Whittelsey Fund, 1962)

3

THE ROCKET IN ASIA

In Asia, as in Europe, preparations for war and a fascination with fireworks provided the twin incentives for the development of rocketry. The origin of the rocket and the nitrate-based powder that makes it function is, however, unclear. Did it emerge in China as widely believed? If so, the rocket could have been carried to the Arab world and to Europe around the middle of the thirteenth century by the westwardly expanding Mongols. Or was the rocket solely, or perhaps independently, discovered in Europe? If solely, post-Marco Polo travelers may have introduced knowledge of the rocket to China in the late thirteenth or early fourteenth century. Or India may have been the birthplace of the rocket, from where the device could have passed on to the Chinese and to the Arabs, after which it made its way into Europe.

Much of the speculation is aggravated by the lack of precise terminology. There is a difference between the discovery and application of simple incendiaries and the controlled combustion of a rocket. In studying Chinese sources, one cannot assume the existence of gunpowder-charged rockets when "old incendiaries" like oil, pitch, and sulfur are meant or such quick-burning mixtures as Greek fire and distilled petroleum and naphtha (none of which were used for propulsive or explosive effects). When one comes to low-nitrate deflagration and high-nitrate explosive mixtures and controlled-burning powders, however, the door is opened for the invention of the rocket. This was as true in China and in India as it was in the Europe of Marcus Graecus, Roger Bacon, and Albertus Magnus.

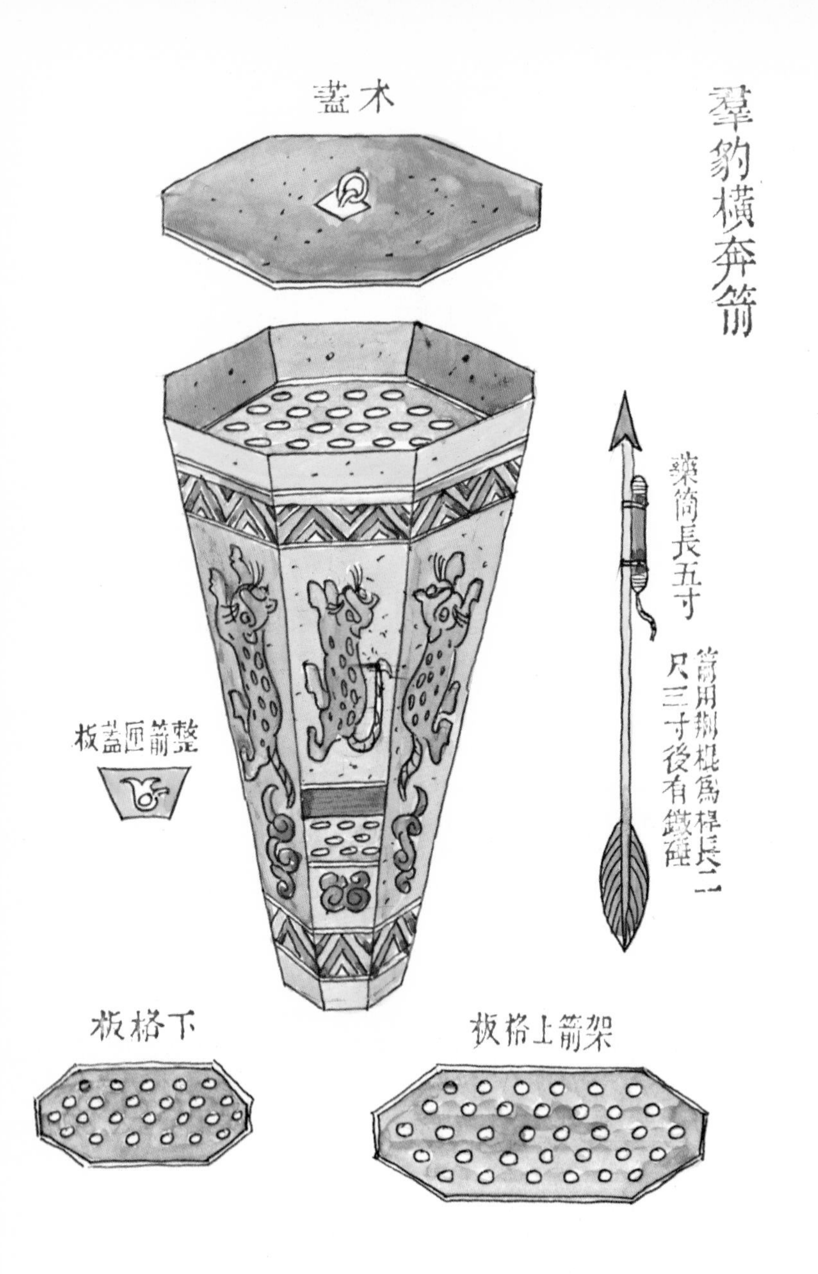

"Leopard-herd-rush-transversally" rocket fire arrow launcher, from the *Wu Pei Chih* of Mao Yuan-I written about A.D. 1621. All rockets in the launcher could be fired at once and could attain distances of up to 400 paces. (Courtesy Harry H.-K. Lange)

Colonel Henry W. L. Hime, in his *Gunpowder and Ammunition* (1904) and *Origin of Artillery* (1915), rejects the belief that the Chinese pioneered in rocketry and denies that they discovered gunpowder. He agrees with W. F. Mayers' "On the Introduction and Use of Gunpowder and Firearms Among the Chinese" (1871 issue of the *Journal of North China Branch of the Royal Asiatic Society*) in which Mayers wrote that every Chinese writer "who treats seriously" the subject of gunpowder disclaims the idea that it was discovered in China. "Every Chinese custom, art, and institution," chided Hime, "is supposed to be very ancient, and what is not really old is readily invested with fictitious antiquity." He also criticized such French sinologists as Antoine Gaubil, Joseph Moriac de Mailla, and Joseph-Maria Amiot for putting too great a faith in the Chinese tracts they studied and translated. In Hime's words, "the Jesuits were blinded by admiration of the Celestials; their critical sagacity was blunted by the air of sincerity displayed in Chinese books."

What did eighteenth-century Jesuits and the Chinese texts have to say? Father Amiot, author of the *Art Militaire des Chinois* (based to a large extent on the A.D. 1621 Chinese book *Wu Pei Chih*, or *Records of War Preparations*), said, in 1772: "I have dealt at length . . . on gunpowder in order to prove that it was known to the ancient Chinese, for everything that I have just reported is taken from their oldest and most authentic works." (He attributed the invention of gunpowder to Sun-Tzu and Wu-Tzu who lived in the fourth century A.D.) The use of gunpowder and firearms since the early part of the Christian era is, he believed, "a fact attested to by all the historians." Moreover, he offered as beyond proof the fact "that the Chinese learned of the propulsive power of gunpowder . . . long before this knowledge had reached Europe." He even explained how to make a powder that would loft a rocket into the skies:

> Add 3/10 of an ounce of sulfur to an ounce of saltpeter . . . and 3/10 of an ounce of charcoal . . . These types of rockets are used for daytime signaling. For nighttime signal rockets, add 2/10 of an ounce of sulfur to four ounces of saltpeter and one ounce of charcoal.

However well he knew the Chinese language, Amiot was

neither a scientist nor a technologist and was forced to leave untranslated many technical terms. "In treating a subject outside his province," observes Partington, "he had opened out a field large enough for his critics . . . to exert all their efforts." Hime had stronger words: Amiot, Mailla, and Gaubil suffered paralysis of "their critical faculty . . . when dealing with Chinese history . . . they evidently did not understand the difference between an explosive and an incendiary."

This failure has long plagued historians, giving rise to many faulty interpretations. How, for example, should one interpret Stanislas Julien's translation of an extract from the *T'ung-Chien-Kang-Mu* relating to the Mongol siege of Pien-king, a town north of the Yellow River now known as Kai-fung-fu?

> At the time [1232] use was made of *huo p'ao* or fire *pao* [incendiary war instrument], called *chen t'ien lei,* or thunder that shakes the heavens. An iron pot was used for that, and was filled with *yo* [the incendiary material]. As soon as it was lit, the *pao* rose up and fire burst forth everywhere. Its noise resembled that of thunder and was heard for a hundred *lis* [thirty-three miles, an impossibility]. It was capable of spreading incendiary materials over an area of half a *mou* [a twelfth of an acre]. The incendiary was even able to pierce iron armor when it attached itself to it.[1]

The fact that something "rose up" had led many to believe that a rocket was involved. Yet, the object is clearly stated to have been an iron pot. When we come to fire that "burst forth," did the author mean a gunpowder-type explosion or simply that the fire spread out from the pot? French military historians Joseph Toussaint Reinaud and Ildephonse Favé suspect the latter:

> The expression "the fire *pao* burst forth" was applied to flashes of flame which spread out through openings [in the pot] and probably should not be taken in the sense of bursting given to our system of artillery. It is not, however, impossible that

[1] (The *T'ung-Chien-Kang-Mu,* a general history of China, is neither the work of a single individual nor of a single period. Rather, parts of it were compiled between 1019 and 1086 by Sua-ma-Kuang under the title *T'ung-Chien,* or *Mirror of History.* He was followed by Chu-hsi (1130–1200), and so on into the seventeenth century.)

> a shattering of the iron pot took place: this would mean the use of a petard, perhaps even of an explosive projectile. (*Journal Asiatique,* October 1849)

Partington agrees: "If the exaggerated language is cut away [that the noise could be heard thirty-three miles away], it is possible to grant that some sort of explosive bomb was concerned."

If so, by the year 1232 the Chinese possessed an explosive powder. But, would they have been able to apply it to the propulsion of rockets? Again, we refer to a translation by Stanislas Julien—or, rather, to our translation from his French:

> Furthermore, the besieged [the Chinese defending Pien-king against Mongol attackers] had at their disposal flying fire arrows (*fei-ho-tsiang*). Attached to the arrow was a material susceptible to catching fire; the arrow flew off abruptly in a straight line, spreading the incendiary material over a distance of ten paces. No one dared approach it [the affected area]. The fire *pao* and the flying fire arrows were greatly feared by the Mongols.

What were the *fei huo ch'iang* (the accepted English, as differentiated from the French, transliteration)? Rockets? Or flying arrows or spears launched by mechanical means? Reinaud and Favé, taking note of the fact that certain Chinese expressions do not always have the same sense and that some words change their meaning over the years, conclude that "this arrow carried an attached rocket near the point; the fuse was lit; the composition enclosed in the rocket took flame; and the arrow was able both to wound and to burn."

Reminding us that the knowledge of making rockets requires a knowledge of preparing purified saltpeter, charcoal, and sulfur, Reinaud and Favé credit the Chinese with being the true pioneers in rocketry. As for Marcus Graecus, who prepared recipes for flying rockets no earlier than 1250, he "very probably had received rudimentary ideas of the Chinese fire arrow." But how? Reinaud and Favé believe it occurred when the Mongols expanded into eastern Europe and the Arab world between approximately 1235 and 1260.

Reinaud and Favé's reasoning does not satisfy Hime, who insists that "Not until we reach the fifteenth century do we meet with gunpowder and cannon." Before that, such incendiaries as were used in China—for example, during the siege of Siang-yang-fu (1268–73)—"were of Western origin and were worked by Western engineers." The trickle of Western influence was strengthened by the residence of the Polos (Marco, his father, and his uncle) from 1273–92 and "was by no means an isolated fact. They were but the pioneers of a considerable body of mechanics, missionaries, and merchants who continued their relations with the country for at least half a century." Hime sums up with the assertion that it was "highly probable that the invention of gunpowder was carried from the West to China, by land or water, at the end of the fourteenth or the beginning of the fifteenth century." He suggests that sixteenth-century Portuguese travelers and merchants, along with Jesuit missionaries who began arriving toward 1600, gave the false impression that gunpowder was an old Chinese invention.

As far as the arrows themselves are concerned, Tenney L. Davis and James R. Ware disagree with Reinaud and Favé's rocket interpretation, preferring to think of the *fei huo ch'iang* as flying fire spears that were "equipped with fire tubes which threw fire forward for a distance of about thirty feet" ("ten paces"). They feel that this was "a reasonable distance for fire to be thrown from a small tube, but it is an unreasonably short trajectory for a rocket and one which would yield but little military advantage" ("Early Chinese Military Pyrotechnics," *Journal of Chemical Education,* November 1947).

Wang Ling, writing in *Isis* in July 1947, offers quite a different interpretation. Impressed by the fact that saltpeter, charcoal, and sulfur—the ingredients of gunpowder—were known "at least as far back as the first century B.C.," he believes that by the eleventh century at the latest "knowledge of the mixture is well established, and practical application of it is being made." He puts the actual discovery of gunpowder "in or before the tenth century," adding that fireworks and possibly rockets were in use in the eleventh century and probably by the middle of the tenth.

> Another type of fire arrow [other than the type with powder placed in the tip for incendiary purposes] is mentioned in the *Wu Ching Tsung Yao:* the *huo yao pien chien,* the description and the name of which clearly indicate the use of gunpowder. It is possible that it was the force of the exploding gunpowder which projected it, as it is stated, that five ounces of gunpowder were applied to the end of the arrow. If so, it was really a rocket weapon.

Research undertaken since 1947 suggests that Wang Ling is correct and that some sort of protogunpowder at least was known in China as far back as the eleventh century. But most modern scholars discount a pre-Christian era date for the powder (as suggested, for example, in a work known as the *Huo Lung Ching* published in 1411) along with Amiot's belief that it was known in China "long before its use in Europe." They do, however, accept the *Wu Ching Tsung Yao* (*Complete Compendium of Military Classics*) edited by Tseng Kung-Liang and published in its original form in 1044. The book describes pyrotechnic materials and devices, including *huo yao* (a term that has been translated both as "devouring fire" and as "incendiary powder"). A later edition of the book mentions "a long bamboo tube filled with explosive power" and a fire lance device that dates from 1132. (Though originally released in 1044, the *Wu Ching Tsung Yao* was up-dated from time to time.) Partington feels that the *huo yao,* which literally means "fire drug," "fire medicine," or "fire chemical," was a low-nitrate powder. Although there is no evidence for it, such a powder could have been applied to rockets. In fact, a thirteenth-century edition of the book states that incendiary arrows were fired in large numbers in the year 1206 with, however, no implication whatsoever that they were rocket-propelled. At the time of original publication, 1044, Partington feels, there existed only protogunpowders which "were used in bombs and not as propellants."

Other Chinese books shed uncertain light on the origin and development of gunpowder in general and rockets in particular. Fire lances (*huo ch'iang*) were used in defending the city of Te-an near Hankow beginning in 1127, according to the late twelfth-century *Shou Ch'eng Lu* by Ch'en Kuei. The dynastic history *Sung Shih* (1345) reports that in 1277 Lou Ch'ien-Hsia ordered the employment of *huo p'ao,* which produced a "clap of thunder" and caused smoke to cover the sky. During 1283 and 1284 the Emperor Kublai Khan's forces invading Japan had fire weapons known as *pao huo p'ao,* but one cannot be sure what they were.

The *Huo Lung Ching* "fire dragon manual" of Liu Chi (Po-Wen, 1311–75) tells us that gunpowder, or something that has been translated as gunpowder, was being made in factories in the fourteenth century. Several thirteenth- and fourteenth-century works describe large quantities of *huo p'ao* being fired. But what was it? The word *p'ao* was sometimes referred to as a projectile, other times to the instrument launching it. The assertion of Lo Ch'i in his *Wu Yuan* that gunpowder fireworks date to the period of the Sui Emperor Yang Ti (603–17) is termed "unbelievable" by Partington. Also rejected is a quote from Fang I-Chih in the encyclopaedia *Wu Li Hsio Shih* that such fireworks go back to the T'ang dynasty, which lasted from A.D. 618 to 906.

Far more important than any of these texts is the *Wu Pei Chih,* completed in 1628 by Mao Yuan-I. Made up of eighty volumes containing 240 chapters, it provides not only recipes for gunpowder and instructions on how to make it, but also gives instructions on how to make several types of rockets and rocket launchers. A so-called "rising powder," presumably used to power rockets, consisted of one ounce of saltpeter, 0.3 ounce of sulfur, 0.04 ounce of litharge (lead monoxide), and between 0.3 and 0.5 ounce of charcoal. Willow, and pine-branch charcoal were preferred in the mixture, which was to be ground in three batches using 5,8co strokes. The artificer was advised to mix the powder with spirits and then shape the result into pellets "about the size of green peas." Moreover, the gunpowder should be used in small quantities "for it is very fast."

One wonders how much Western influence was exerted on the author of *Wu Pei Chih.* Beginning about 1250 some communication existed between China and the West, and, from the early 1400s, between China and the west coast of India. By the sixteenth century both the Portuguese and the French Jesuits were in contact with the Chinese, a fact that affected the subsequent development of firearms in China. Not only are Portuguese firearms described by Mao Yuan-I but, as Partington points out, "there were always Europeans able and willing to impart military information for suitable reward." For their part, Jesuits assisted the Chinese in learning how to cast cannons. "In my opinion," Partington concludes, "it would be very unsafe to assume that Mao Yuan-I, a very well-informed man who had been a military commander, was ignorant of the composition of European gunpowder or of European firearms." If this is true, then the advance from the protogunpowder used—apparently—by the Chinese since the eleventh century may well have been influenced by the introduction of European gunpowder in the sixteenth and seventeenth centuries.

A number of rockets are illustrated in the *Wu Pei Chih,* the most important being fire arrows with attached rocket charges and guiding sticks. These were ignited and launched from a kind of crossbow, the rocket power being relied upon to give them additional range. Basket-shaped and box-shaped launchers are pictured, among them a device from which arrows "will rush out on a solid front like a hundred tigers." Davis and Ware translated the Chinese text:

> The powder capsule [drive tube] is 0.3 feet long. For the arrow shaft, employ bamboo 1.6 feet long. Behind the feathers there is an iron weight . . . Let there be 100 arrows. Behind the feather place an iron weight so that the center of gravity will be four fingers [width] from the mouth of the capsule. It will be possible to shoot over 300 paces, 100 arrows at a time. If smeared with tiger-shooting poison, the might [of the arrows] will be very fierce, whence the name ["fierce" of the arrows].

Despite continuing research, it remains unclear whether rockets appeared in China before they did in Europe. Partington thinks that gunpowder, "of more or less modern composition," was known to the Chinese by the end of the Mongol Yuan dynasty, which lasted from 1260 to 1368. "It is uncertain whether it [gunpowder] was developed by the Chinese or the Mongols, or

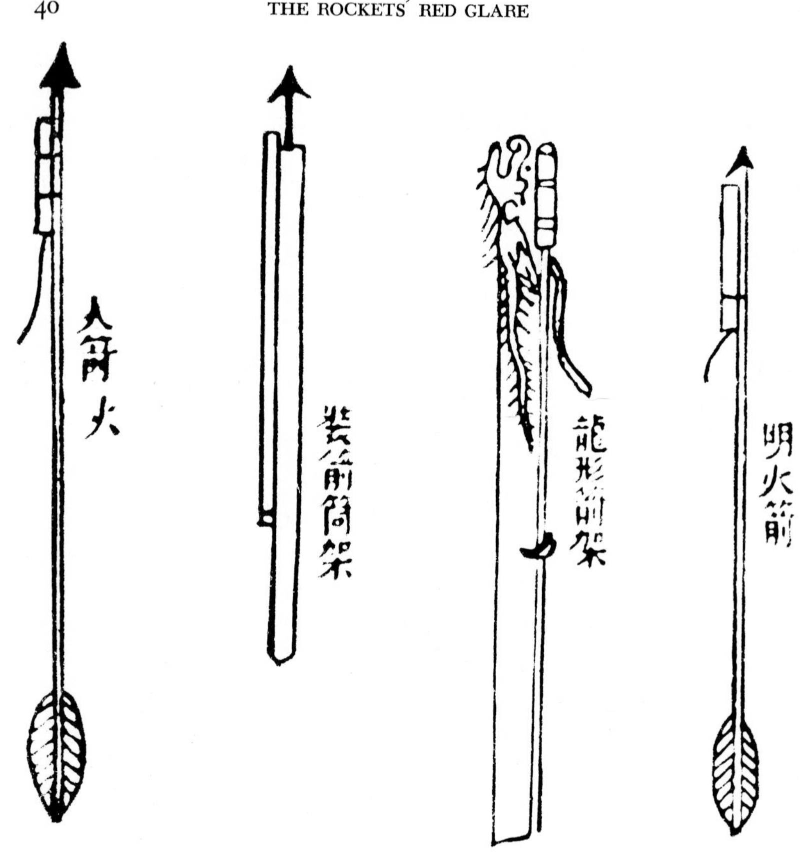

Rocket-launching tube and rocket-powered fire arrow from the *Wu Pei Chih.* (Courtesy James R. Ware)

even if knowledge of it came from the West," he adds. He suspects it was discovered in China. L. Carrington Goodrich and Fêng Chia-Shêng, writing in *Isis* in January 1946, believe:

> there is valid literary evidence for the development, by the thirteenth century, of real firearms in China, although dated examples are lacking until 1356 . . . The literature indicates that around the year 1000 the Chinese had flame-throwing devices. By 1132 they were using long bamboo tubes filled

> with explosive powder, and by 1259 bullets were inserted in these tubes and ejected by touching off the powder.

They do not speculate when the rocket was introduced.

However one may wish to interpret the Chinese documentation, it is very hard to make a case for the appearance of rockets before the thirteenth century—or about the time they were being described by Arab and European writers. Sooner or later, evidence may indicate a Chinese origin; and, in fact, the studies of Joseph Needham and his associates at Cambridge University show that "we have to consider the machines or devices . . . which had reached Europe from China at earlier times, or were still to be transferred there [to Europe]. Among these characteristically Chinese inventions we may list . . . rocket flight." In Volume 4, Part III of *Science and Civilization in China,* a passage from the *Sung Hiu Yao Kao* (*Drafts for the History of the Administrative Statutes of the Sung Dynasty*) is quoted, stating that *hui chien,* or incendiary arrows, "are at this date [1124] very probably rockets."

Although fireworks got an early start in China, the art did not develop as rapidly there as in Europe. That an expert like Alan Brock was not overly impressed with them is evident: "Chinese firework displays have often been enthusiastically described by travellers in China. Whether it is that the glamour of the East distorts the perceptions, or that these travellers have not seen a European firework display, there is no doubt that such descriptions are, to say the least, overcoloured." Despite earlier Chinese dominance, the Japanese soon became Asia's leading pyrotechnicians, the effects of their aerial firework displays reaching "a remarkable degree of perfection."

The Indians, too, became very adept at making fireworks and soon surpassed the Chinese. However, there is no evidence as to whether or not fireworks were invented in India. Much of what is known about them was gathered by P. K. Gode, who before he died was curator of the Bhandarkar Oriental Research Institute in Poona.

The earliest Gode could date fireworks from the written record was the middle of the fifteenth century. The first evidence,

Japanese prints of fireworks manufactured by Hirayama of Yokohama, late nineteenth century.

published by the Indian Institute of Culture, comes from a communication by Abdur Razzāq, ambassador from the court of Sultan Shāh Rukh of Persia, who stayed in Vijayanagar from late April to early December 1443. He is quoted as saying, "One cannot without entering into a great detail mention all the various kinds of pyrotechny and squibs and various other amusements which were exhibited." Just how long before 1443 fireworks were used in India it is impossible to say; but, since saltpeter, charcoal, and sulfur are common in the country, it could well have been many years, even centuries.

Other evidence shows that fireworks were flourishing in Kashmir and as far south as Sumatra and Malacca, by 1500, "if not earlier." Typical extracts from contemporary writings are the following:

> [Re elephants] But if at any time they [the elephants] are bent on flight, it is impossible to restrain them; for this race of people [the citizens of Vijayanagar] are great masters of the art of making fireworks and these animals have a great dread of fire, and through this means they sometimes take to flight.

> [Re a Brahmin wedding in Gujarat] During this time they [the bride and bridegroom] are entertained by the people with dances and songs, firing of bombs and rockets in plenty, for their pleasure.

In the manuscript *Kaututkacintāmaṇi* by Gajapati Pratāparudradeva of Orissa (1497–1539), Gode found Sanskrit verses containing formulas for preparing fireworks of sulfur, saltpeter, charcoal, powder of steel, and powder of iron. The listing of these ingredients around 1500 "speaks for itself," remarks Gode. "It is possible," he continues,

> to suggest that the Chinese formulas for the manufacture of fireworks were brought to India some time about A.D. 1400 and then modified by the use of Indian substitutes for Chinese ingredients, not all of which may have been then available in India [Gode is referring to coloring and other secondary materials]. The main pyrotechnic ingredients, like sulphur, saltpetre, charcoal, powder of iron, etc. had to be retained in the Indian formulas as they were the very basis of pyrotechny; they were available in India from early times.

Another important discovery from the Sanskrit literature is the description of structures used in pyrotechnic displays prepared for royal entertainment. The *Ākāśabhairava-Kalpa* tells how *bāṇas,* or rockets, were hung from these structures and made ready for firing and how later one rocket kept in reserve would be set off to indicate the end of the spectacle. The word *bāṇa,* incidentally, in the sense of its being used to mean a rocket, is not of Sanskrit origin; its history and etymology remain to be traced.

Firework making and displays in India continued uninterrupted throughout the seventeenth, eighteenth, and nineteenth centuries. Toward the end of the eighteenth century, the English added their talents to those of the Indians, offering their own displays "to please Indian princes by their skill in the art of fireworks" (Rao Bahadur D. B. Parasnis, "English Fireworks in India," in *Itahāsa-Saṁgraha,* January 1909). From the description of the work of an Englishman who called himself "Karar," we learn that "As the display was coming to a close there arose from the blazing fire of a rocket a Sun of fire, which moved high in the sky . . . This was followed by another rocket producing a similar display of the Moon."

Just as it is difficult to authenticate the use of fireworks in India before A.D. 1400, so it is with the military applications of protogunpowder and gunpowder itself. Modern scholars do not accept such statements as, "Gunpowder and firearms were known in India in the most ancient times . . . the knowledge of making gunpowder was never forgotten in India . . ." (Quentin Craufurd, *Sketches Chiefly Relating to the History, Religion, Learning, and Manners of the Hindoos,* London, 1790). How far back, then, can one safely go?

Only to the very end of the fourteenth century.

One bit of evidence for the use of fire weapons and rockets comes from Sharaf al-dīn 'Alī Yazdi's *Zefār-Nāma,* the Persian history of Tamerlaine (Tīmūr). During an engagement near Delhi in 1399 against the defending Sultan Maḥmud, "flying rockets with iron tips" were employed against Tamerlaine with good results. Tamerlaine (c. 1336–1405) says in his *Autobiography* (*Malfūẓāt-i-Tīmūrī*) that he was opposed by grenade throwers, fireworks, and rockets (*takhshandāzān*). Although Tamerlaine's forces did not enjoy such weapons, they carried the day with arrows and swords.

Subsequent Indian history is sprinkled with accounts of military

A rocket soldier in the army of Tippoo Sahib during the final decade of the eighteenth century in India. Notice how the rocket is lashed to the long bamboo stick.

rockets. They played an important role when the forces under the Mogul Emperor Aurangzeb captured the fort of Bitar in 1657. According to Sir H. M. Elliot (*The History of India as Told by Its Own Historians: The Muhammadan Period,* 1867), one of the defenders' rockets flew accidentally into their own munitions storage room creating such commotion and casualties that surrender followed promptly. Events such as this, however spectacular they may have been, were sporadic. It would not be until the latter half of the eighteenth century that the Indian war rocket would make its mark on history.

The major European maritime nations were involved commercially and militarily in India from the arrival of the Portuguese in 1498 to the division of the subcontinent into India and Pakistan in 1947. For a hundred years after they defeated an Arab-Egyptian fleet in 1509, the Portuguese dominated not only the Indian Ocean but European commerce in India as well. Then came the British and their East India Company (which had been set up in 1600 as the Company of Merchants of London Trading with the East Indies), the Dutch (whose United East India Company of the Netherlands was established two years later), and the French (in 1664, Louis XIV ordered the creation of the Compagnie des Indes Orientales). Inevitably, the interest of these powers clashed, and by the mid-eighteenth century Portuguese and Dutch influence had waned and the British and French—with the Indian princes they each supported—found themselves in an almost continuous state of conflict.

As early as 1750, a French army under Colonel Charles de Bussy came into contact with an Indian war rocket. On August 21 of that year, during an engagement with the forces of Mohammed Ali in the Carnatic, the Indians let loose a single rocket that terrified the French horses and set fire to an ammunition cart. It, in turn, injured several soldiers. Six years later, the Nawab Siraj-ud-daula suddenly captured the British settlement at Calcutta. During the attack, his troops loosed incendiary rockets against English ships in the harbor in hopes of setting fire to them. Still later, in 1781, Prince Haidar Ali of Mysore invaded the Carnatic, engaging the forces of Colonel Eyre Coote. During the fighting, some 1,200 Mysore rocket soldiers became involved. In one battle, on September 9 and 10, a contemporary history of Haidar Ali trumpeted that Indian rockets "ravaged murderously" the English lines, exploding some ammunition carts and causing a complete rout. During a withdrawal action in 1783, Innes Munro told how the English "were constantly pursued by a strong squadron and rocket-throwers, so that the English rear-guard lost 200 men" (*A Narrative of the Military*

Operations on the Coromandel Coast Against the Combined Forces of the French, Dutch and Hyder Ally . . . , 1789).

Haidar Ali's successor, Prince Tippoo Sahib of Mysore, not only continued to use the rocket forces as he fought against the British, but increased their strength to 5,000 men. He used them effectively in 1790 during an unprovoked attack on Travancore, which was allied to the English; and again in 1792 against Lord Cornwallis during the siege of Seringapatam. Rockets notwithstanding, the Prince of Mysore lost the battle and was obliged to yield half his territories and pay a large indemnity. He persisted in intriguing against the British, however. With the aid of a French force moved up from Mautitius, Tippoo Sahib again went to war in 1799. Again his rocket forces played their part. "So pestered were we with the rocket," muttered a young British officer, "that there was no moving without danger from the destructive missiles." After three months of hostilities, Tippoo was finally defeated and killed in May 1799.

His war rockets were simple, easily carried, not very accurate, cheap to make, uncomplicated to operate, and readily produced. One type consisted of a 10-inch long, 2½-inch exterior diameter iron case lashed to a sword blade by strips of hide. Another was just under 8 feet in length and 1½ inches in exterior diameter; a 6¼-foot long bamboo stick was substituted for the sword. Still other rockets were lashed to bamboo sticks up to 10 feet in length. All the tubes had sharp tips to assist in penetrating the targets. Depending on the nature of the battleground, the Mysorian soldiers would fire their rockets along a conventional aerial trajectory, which might be a thousand yards long, or horizontally along the ground. Though a single rocket could kill or injure up to six men, the weapon was most effective against horses and in setting off ammunition carts.

Although there is no documented connection between the development of protogunpowder, gunpowder, and rockets in China and in India, the technology may have passed from the former to the latter, or even the other way around; Partington does not discount the possibility that "knowledge of gunpowder may have reached China from India." Nor is there any evidence of pre-Portuguese (i.e., pre-1498) Western influence on Indian developments. But what about the Arab world? Did the Arabs learn about gunpowder on their own? Or from the Chinese via the Mongols who invaded Persia, Iraq, Asia Minor and Syria in the mid-thirteenth century? If the Indians knew about gunpowder before then, *they* may have been the first to pass on the discovery to the Arabs; or, of course, the Arabs could have acquired European gunpowder and rocket recipes during the course of the Crusades, which began at the end of the eleventh century.

How far back can gunpowder be dated in the Arab world? And when was it (or a low nitrate protogunpowder) applied to rockets? These are not easy questions to answer.

The Arabs fired incendiary arrows during the first invasion of India in A.D. 712 and continued employing these and other incendiary devices during their entire period of expansion eastward and westward. They used them during the Crusades from the end of the eleventh century through the fourteenth. While it is unclear when the explosive and propulsive applications of protogunpowder and gunpowder were introduced, it was probably not before the mid-thirteenth century. By the fourteenth century the documentation is unquestioned. During the Spanish siege of Algeciras (1342–44), for example, Juan de Mariana (1536–1623) writes, "The besieged [the Arabs] did great harm among the Christians with iron bullets that they shot. This is the first time we find any mention of gunpowder and ball in our histories [e.g., the history of Spain]." Similarly, Arab artillery was deployed in Egypt and in Syria from at least 1366. Both these mid-fourteenth-century dates are about a century after Marcus Graecus, Roger Bacon, and Albertus Magnus, meaning that the Arabs may have learned about gunpowder from Europe.

What do earlier Arab manuscripts tell us, particularly equivalents—in part at least—of *Liber Ignium*? They record fire arrows and various incendiary compositions that could be thrown at, or hurled upon, the enemy. Two important manuscripts, reposing in the Leyden (Holland) library, of the original *Treatise on Strategems, Wars, the Capture of Towns, and the Defense of Passes . . .* (composed around 1225) fail, however, to mention gunpowder, let alone its critical component saltpeter. In the opinion of Reinaud and Favé, "the various incendiary compositions used by the Arabs and by the Greeks before the year 1225 do not contain saltpeter."

By the end of the thirteenth century, however, they do.

Gunpowder and saltpeter are described in an extremely important manuscript by the Syrian Al-Ḥasan al-Rammāh (the Lancer) Nadjm al-dīn (the Star of Religion) al-Aḥdab (the Hunchback). Written between about 1285 and 1295, the year of his death, it is entitled *Treatise on Horsemanship and Strategems of War.* It describes how to purify saltpeter and how to make all manner of pyrotechnic devices, including rockets for which several compositions are given (one calls for 10 parts saltpeter, 1½ parts sulfur, and 3 parts charcoal, while the ratio for another is 10 to 1¼ to 2½). Chinese influence is apparent throughout the manuscript, which was written decades after the Mongol invasions of the Near East. (Baghdad, for instance, was sacked in 1258, and in 1259 the armies of the Mongol ruler Hulagu moved into Syria capturing Damascus and Aleppo the following year.)

One of the most intriguing rocket-propelled devices in the Ḥasan al-Rammāh manuscript is "the egg that moves itself and burns," a pear-shaped, torpedolike object with two protruding rockets that may have helped guide the device and a larger centrally mounted rocket that presumably provided the main power. The "egg" was used as early as 1248, for the Sire de Joinville, in his *Histoire de Roy Saint Loys,* reports that French soldiers operating near Damietta were suddenly confronted with "a projectile . . . which, when it had fallen on the bank [the river] came straight towards them, burning wildly; it is doubtless the egg that moves and burns."

Still another manuscript, the *Collection Combining the Various Branches of the Art,* has been studied to see what light it may shed on Arabic developments of gunpowder. Obtained by the Count of Rzevuski and later acquired by the Saint Petersburg Academy—and hence known as the Saint Petersburg manuscript—it was probably composed between the years 1300 and 1350 by Shams al-dīn Muhammad. Of particular interest is the description of an "arrow from Cathay," which Reinaud and Favé

By the end of the nineteenth century, Asia dominated the firework rocket scene. Here an aerial burst lights up the night sky.

assume to be a rocket. Their translation of the ingredients: "saltpeter, 10; charcoal, 2⅝ drachmas [approximately the same as the apothecary's dram]; sulfur, 1¼ drachmas." The manuscript explains how the powder is pulverized and loaded into the case. The rocket is then attached to a lance and the fuse is placed "at the opposite end."

As a result of their studies of the Leyden manuscript, that of Ḥasan al-Rammāh at the Bibliothèque Nationale in Paris, the Saint Petersburg manuscript and others, as well as evidence from Europe and from China, Reinaud and Favé believe that knowledge of rocket-propelled fire arrows "penetrated the Arab world and the Western Christian world on the heels of the Mongol armies around the middle of the thirteenth century." They credit the Chinese as being the first to purify saltpeter and to use it in fireworks. "They were the first to mix this substance with sulfur and charcoal," they conclude, "and to recognize the propelling force that is born from the burning mixture. This is what gave them the idea of the rocket."

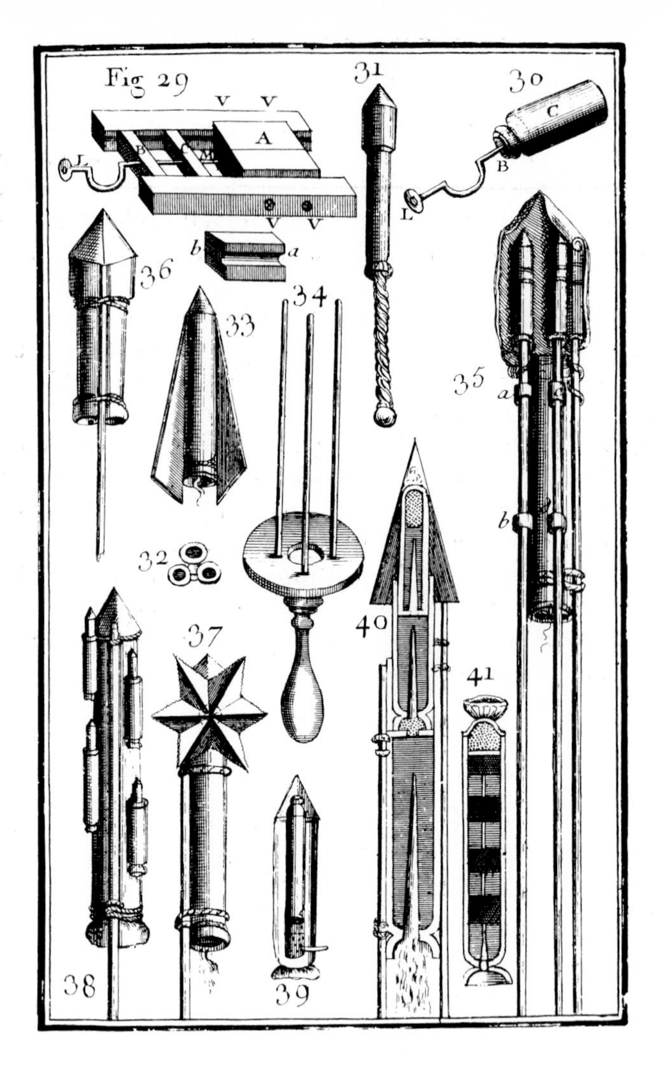

This plate, reproduced from Amédée François Frezier's 1747 *Traité des Feux d'Artifice*, illustrates a rocket being guided by suspending a weight at the end of a cord instead of using the conventional guide stick (31); a cluster of three rockets shown in cross section (32) and side view (36); ailerons added to provide stability (33); a hand-held launcher for these rockets (34); a two-stage rocket (35); a rocket with a large star as its payload (37); a rocket that releases small rockets as it rises (38); a more detailed view of one of the small rockets (39) released from the large one; a three-stage rocket with delta-winged third stage (40); and the cross section of a rocket named "Fury" for "its effects are sometime lively and sometime slow" (41).

4

THE OPERATION AND MAKING OF ROCKETS

Until the twentieth century, rockets were made, loaded, and fired by hand. A. St. H. Brock observed in 1922 that "The methods of charging rockets in use in the sixteenth century are those of today, and it is remarkable that no satisfactory alternative to hand charging has yet been divised." The mere description of a rocket by, say, Mortimer in 1824 is that of a Bate, Babington, Malthus or Appier (Hanzelet) in the early sixteenth century or of a Brock in the early twentieth:

> Rockets consist of strong paper cylinders, which being filled with the proper composition rammed hard, and fire being applied to their apertures they are caused to ascend into the air, or in any required directions; they have generally a head fixed to them containing corn powder, sparks, and many other decorations, which, when the body of the Rocket is consumed, take fire, burst in the air, and produce a most beautiful appearance. (G. W. Mortimer, *A Manual of Pyrotechny*, 1824)

James Cutbush (*A System of Pyrotechny*, 1825) added another essential: "It [the rocket] is furnished with a stick, serving as a counter-weight, or balance, to guide it vertically in its ascension." He also noted that some rockets, e.g., those used by the military, were routinely fitted with iron cases and iron conical heads and contained shot or incendiary mixtures designed to cause damage to an enemy.

Issac Newton expressed his famous third law of motion thus: for every action there is an equal and opposite reaction (*Philo-*

The mixing of gunpowder is shown here with the following description: "If thou wishest to make a good strong powder take four lbs. of saltpetre, and of sulphur one, and one of charcoal, and one ounce of salpratica, and one ounce of salammoniac and 1/12 part of camphor and pound all well together, and add burnt wine thereto and mix it therewith and dry it well in the sun, thus shalt thus have as strong a powder as thou couldst wish; one pound of it will do more than three pounds else, and it will keep well and improve with age." From the *Codex Germanicus* No. 600, c. 1350.

sophiae maturalis principia mathematica, 1687). When the powder in the rocket chamber is ignited, combustion commences which in turn produces hot, pressurized gases. These gases act to exert force equally in all directions; but, because of the opening at the throat end of the rocket, there is an unbalance of force that is proportional to the chamber pressure and to the throat area. This is responsible for the forward motion of the rocket. As the back pressure of the outside air *decreases,* the unbalance *increases.* This means simply that a rocket functions better in a vacuum than in the atmosphere and explains why the early space enthusiasts became so attached to the device. Few of the early rocketeers understood this, feeling rather that the exhausting gases somehow pushed or reacted against the air, making the rocket projectile fly in the opposite direction.

Master pyrotechnician Claude-Fortuné Ruggieri, writing in 1812 on the cause of rocket ascent in *Pyrotechnic Militaire,* recognized that "the ascent of a rocket is due to the expulsion of gases and heat," but erred when he stated that the gases "lean on the air." James Cutbush put it this way in 1825: "The gases proceeding from the interior of the rocket, act therefore upon the air in the immediate vicinity of the orifice, and the rocket is consequently propelled, the stick guiding it in the direction given to it." Cutbush was aware of the reaction principle, but did not fully understand its application to rocketry. "As action and reaction must be equal," he wrote, "the repulsion produced by the action of the gases upon the air is equal to the force impelling the rocket." The air, of course, only impedes a rocket, not impels it.

G. W. Mortimer felt that "This ascent [of a rocket] is dependent on the medium (or air) in which it is generated . . ." Even Henry B. Faber, dean of the Pyrotechnic School, U. S. Army Ordnance Department, theorized in 1919 that a rocket is a "projectile containing a composition which, as it burns, generates sufficient gas to drive the rocket forward by reaction against the inertia of the air. As the gas escapes at the base of the rocket, it encounters the resistance of the air, and in recoil from this the rocket is itself forced upward." Again, the correct explanation was tantalizingly close, but for most people it was hard to resist

Late sixteenth-century illustration of a liquor boiling room in a saltpeter refinery according to manuscript by Lazarus Ercker, *Beschreibung aller fürnehmsten mineralischen Ertzt,* or *Description of Principal Precious Minerals.* After having been scraped from walls or artificially prepared in heaps, saltpeter was dissolved in water and then boiled. The scum is removed from the vat and the liquor then allowed to purify until clear. To this end, various purifying agents were added.

the idea that the outside air was crucial to a rocket's performance. For most, but not for all: William Hale in 1863 (see Chapter 5) saw correctly that "it is right to conclude that the effect of the air upon the escaping gas is of no avail in propelling the rocket."

Whether or not the pyrotechnicians and artificers understood the theoretical basis for rocket motion, they knew how to make rockets. First, the rockets were cylindrical and usually were formed of pasteboard—though for war purposes their cases were almost always made of iron. Typically, the outer diameter of the cylinder was between 1½ and 2 inches, while the length of the propellant powder charge measured five diameters. The interior diameter was two thirds that of the exterior diameter.

The composition of the powder charge varied little from the sixteenth through the nineteenth centuries. Saltpeter served as the oxidizer and was purified by being melted in water and then boiled until a film appeared. While the main mass was thickening, some alum was added giving rise to a thick scum that had to be skimmed off. On the third day, all the water was removed. After further purification, the saltpeter was ground in a mortar or else crushed on a table with a wooden pestle. Finally, it was pulverized by passage through a silk screen and left to dry.

The second ingredient, sulfur, also had to be purified by being melted and strained through a cloth. This done, the pure product was reduced to a powder by passing it through a small-mesh silk screen. As for the charcoal, the final ingredient, artificers usually preferred soft, light woods such as willow as the raw material. First, the wood had to be cleaned of its bark and then dried. Next, it was placed in a fireplace or stove for burning. As soon as embers appeared, they were extinguished and the resulting cinder were transferred to a sieve. All three ingredients could then be mixed in the desired proportions, which depended on the purpose. If the powder was to explode in a firework display it was prepared as a grain, but if it was to propel a rocket it had to be pulverized. The pulverization process took place on a wooden table, the artificer doing the crushing with a wooden pestle. This accomplished, the powder was passed through a fine-mesh silk screen, then dried, and finally stored

"Wilt thou see if sulphur be good or not, then take a lump in thy hands and lift it to thine ears. If the sulphur crackle . . . it is good, but if sulphur keep silent and do not crackle, it is not good and must be treated according to what thou shalt hear later on how it should be treated. Sulphur not containing more than a per cent of impurities does not crackle." From the *Codex Germanicus* No. 600.

in canvas bags packed in barrels where they would remain until ready for charging (loading) into a rocket case.

The inner part of pasteboard case rockets was made up of several sheets of good-grade paper, while the outer part consisted of plain white paper. Wheat flour paste was the adhering agent. Small rockets were rolled from three layers of pasteboard, while five sheets were used for large rockets. Sheets were prepared by hand and left for six hours in a press or flattened down by boards on top of which weights were placed. The pressed pasteboard was then hung to dry, sometimes being returned to the press to remove any warps that may have formed.

Two other elements remained: the fuse and the priming. The fuse, used to ignite the rocket, was made of cotton thread soaked in vinegar or brandy for several hours. It was completed by working in powder that had been sprinkled over the thread. It was left to dry and then was stored. A typical fuse was 2½ feet long. The priming, produced by moistening grain powder into a paste, served to seal the fuse in the orifice of the rocket.

Once the rocket case and the powder composition had been prepared, the loading or charging operation could begin. In order to hold the case firmly during powder loading, a hollow, round wooden or metal mold sitting on a cylindrical base was employed. Running upwards from the middle was a spindle or piercer, used to "maintain a cavity within the combustible material or charge" (Frezier). This cavity or conical hole was called the "soul" (*âme*) by the French artificers, for it was where the combustion gases were created and where the reactive forces were generated.

After a case had been made by rolling pasteboard over a *baguette* (round piece of rolling wood or stick), the orifice area had to be choked in. This was done before the case was completely dry by pulling taut a cord over the pasteboard to the diameter of a small stick used to maintain the desired opening. The resultant hole just permitted the piercer to enter; to prevent the hole from loosening up, an "artificer's knot" of thread was tied around the neck (Frezier described it as being "like a collar and a tie").

The case made and choking accomplished, the delicate load-

Quality charcoal made from selected woods, including twigs, was indispensable to the making of first-rate gunpowder. One manuscript exhorted: "Thou shalt make good charcoal. Take wood of the lime or poplar, that is the best, and roast it well in a baking oven, and burn it wholly and entirely, and put of this a goodly quantity into a basin, and turn another basin over it and so close up the charcoal." To make the very best, the manuscript continues, "slack [slake] it with burnt wine or else with good wine and dry the charcoal in the sun. Thus thou shalt have good charcoal." Here, in a sixteenth-century illustration, the lid of a wooden cask is being taken off and a liquid in the flagon will be poured in to slake the charcoal. The pieces suspended in the background are beechwood fungus, used (when dried) as tinder.

ing or charging operation could begin. Among the instruments developed by the French artificers were the *cornée* (loading spoon) and a number of hollowed-out loading sticks. When the case had been placed in the mold in its correct position over the piercer, the first loading stick was placed in the case and then hit gently with a hardwood mallet. The stick was then removed and the powder poured in. Again, a loading stick was inserted and the mallet brought into play to compress the powder. Successive loading sticks were used as the powder built up, each loading operation requiring five less taps with the mallet than the one immediately preceding. Nine to ten charges covered the piercer, with several more needed for the portion above. (This piercer was called the *massif* since it was entirely solid.)

Once the charge had attained the level of the mold, some paper was placed on it (making a wadding) and tapped firmly in place. Then the remaining part of the case was folded over this wadding and again the mallet was used to tap it gently. This completed, the rocket was taken off the piercer and out of the mold, and the folded pasteboard was trimmed. If the rocket was not to be fired shortly after manufacture, it was protected by pasting paper over each end and then was stored in a dry environment.

When being made ready for firing, guiding sticks made of pine, elm, willow, or walnut wood had to be added. Artificer Perrinet d'Orval explained, in his *Essay sur les Feux d'Artifice . . .* (1745), that "They should be at least nine times longer than the rocket itself, not including the *garniture* [fireworks attachments], whose height varied." Moreover, the thickest part of the stick, the part attached to the rocket case, "should only have half the exterior diameter of the rocket, at the most." The stick gradually tapered until it became an eighth of the exterior diameter of the rocket. Perrinet d'Orval added that "The longer the guiding stick, the straighter the rocket would rise." Grooves had to be notched in the stick to enable it to be attached to the rocket; the attachment points were located below the head end of the rocket as well as at the choked-in orifice end. Occasionally, small lead balls suspended by two- to three-foot-long wire were used by nineteenth-century artificers to "balance" rockets. This technique was not, however, successful and was soon abandoned.

Upper left: The choking operation. Choking should always take place *before* the case is dry. Here, the cord is seen attached to a loop or staple driven into the post while the other end is fastened to a length of wood held between artificer's legs. By leaning backwards, his weight enables him to do the choking. Care must be taken to revolve the case so that the choking is accomplished evenly around the throat, leaving a hole barely large enough to permit the piercer to enter. As soon as the choking is completed, the case is tied by passing around it three nooses of thread and tightening each loop in a knot. Upper right: The charging or loading operation. The principal instrument is the hardwood mallet. Below: Cases are made by rolling pasteboard over the *baguette*, or cylindrical-shaped length of wood. Once the case is shaped and rolled, it is made ready for choking. According to Perrinet d'Orval, the pasteboard is gray on one side and white on the other. French print of 1750.

Rocket launchers were relatively simple, usually consisting of a rack or post, or a V-shaped trough. In the case of the former, the rocket would hang on a screw, nail, or hook at the junction point of the throat and the guiding stick. Frezier describes a multiple launching frame from which a number of rockets could be fired almost simultaneously, as well as a post permitting rockets to be launched various degrees away from the vertical. He even mentions a sextant that permitted one to "compare the heights reached by the rockets and then judge from that the force of their composition."

Cutbush and others objected to the method of hanging rockets on hooks or nails, preferring rather to use a ring made of strong iron wire "large enough for the stick to go in as far as the mouth of the rockets."

> Then [Cutbush continued] let this ring be supported by a small iron, at some distance from the post or stand to which it is fixed; and have another ring fixed in the same manner to receive and guide the small end of the stick. Rockets thus suspended will have nothing to obstruct their flight. The upright, to which the rings are fixed by the small iron, must be exactly vertical.

Tubular launchers became popular in the nineteenth century as they could be raised and depressed and swung left or right in accordance with target distance and wind condition. Some military rockets were launched along the ground in large numbers in what was termed a "ground volley." They would slither across a field—which had to be quite flat—for 100 to 150 yards, then they might rise up slightly and rush around in what one witness called "a most destructive manner."

Dealing with explosive mixtures, artificers were constantly exposed to accidents that ranged from slight to lethal. Amédée Denisse estimated, in his *Traité Pratique Complet des Feux d'Artifice* (1882), that "nine times out of ten, accidents occur in the shop where loading [of the powder] takes place" and not—as one might expect—at the time of actual firing. The most common cause, he reported, was the "presence of grains of sand" in the powder composition, particularly if they contained mica. When only small quantities of powder were inadvertently ignited, the accident need not be deadly, ordinarily causing "only light burns that, nevertheless, require immediate care and several days of rest." Should, however, a large amount of powder be present when an explosion occurs, the entire manufacturing establishment might be annihilated. "It is thus necessary to insist," wrote Denisse, "that loaders only work with small quantities of powder at a time, and that they take their tubes [rocket cases], the moment they are loaded, to another room in the shop."

The history of rocketry is filled with reports of accidents. Brock, "struck by the frequency with which explosions occur as a result of ignorance, generally on the part of amateur firework makers," was equally concerned about the plight of the professional. A typical accident reported by Brock took place in 1885 at Mitcham:

> The cause of this occurrence was quite simple. A man was fixing the curved stick which forms the pivot upon which a tourbillion [a type of firework rocket] rotates to one of those fireworks. The wire nail used for the purpose penetrated the composition and fired it. The remaining goods in the shed were ignited and communicated to the neighbouring buildings, one of which was a magazine containing 3,000 lbs. of partially manufactured fireworks, including a number of rockets. These being without sticks and becoming ignited flew in all directions, setting fire to other buildings. The result was that ten buildings and an air drying rack were totally destroyed, and three buildings and three racks partially so.

Brock noted that between 1891 and 1894 alone, thirteen fireworks factory accidents occurred in England and eight in the United States.

A century and a half earlier, an ever-cautious Frezier offered a remedy against firework burns:

> Boil wholesome and fresh pork [fat] in common water using a moderate flame. Then remove it and expose it to evening dew for three or four nights, after which place it in an earthen

> pot and let it melt over a low fire. When melted, run it across a cloth soaked in cold water. Wash it [the pork fat] several times with clear, fresh water until it becomes white as snow. Then place it in an earthen vessel until the occasion arises when you may need it.

Applying the potion was simple. All you had to do was to anoint the burned skin with as much as needed "and you will soon observe an admirable effect."

5

KINDS OF ROCKETS

Over the centuries, rockets came to be designated by their exterior diameters, interior case diameters, and the weight of a lead ball that was just able to fit into the mold in which the powder loading took place. Propellant composition also varied with use; larger rockets required a larger proportion of sulfur and charcoal than smaller ones. Increasing the proportion of charcoal in a given rocket produced a brighter, longer exhaust tail, useful for fireworks and for signals. Many artificers added iron filings to produce spectacular burning effects.

Most commonly flown were simple firework rockets, or skyrockets, which were stick guided and fitted with a head, or pot, that contained the garniture (the term "furniture" was also used). This contained an explosive mixture designed to produce a loud clap, or perhaps a composition that would create falling stars, or a charge of picrate of potash to produce a whistling effect. Or some other "payload." In his *Manual of Pyrotechny*, G. W. Mortimer explains the purpose of one part of a skyrocket's payload, the petard. This, he wrote,

> is a small round box of tin-plate united to the diameter of the [rocket] case, and filled with fine gunpowder; it is deposited on the composition after the ramming, and the remaining paper folded down over it to keep it secure; the petard produces its effect when the Rocket is in the air and the composition is consumed.

Other garniture could be attached, for example, adjusting to

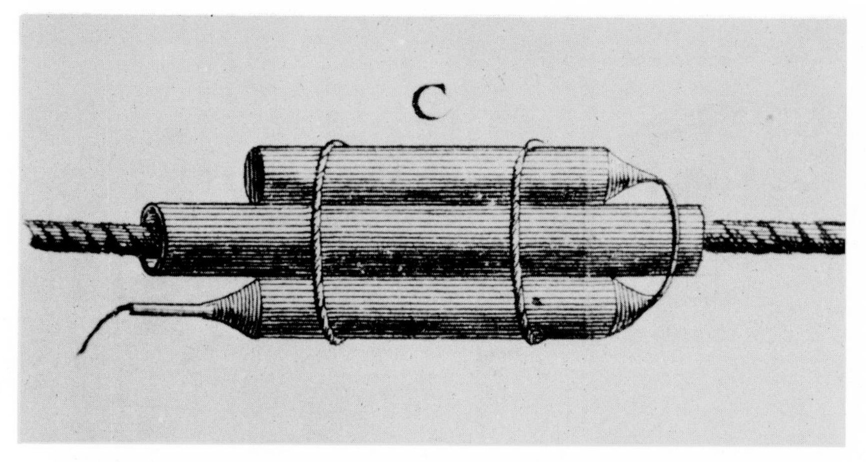

The double *courantin,* or line rocket, attached to a cord, using an empty cartridge as the runner. This design dates from the eighteenth century.

a rocket's head "an empty pot or cartridge of larger dimensions than itself, in order that it may contain the various appendages, which are to render it so superior to the others, in the beauty and splendour of its emication [sparkling]."

An interesting variation of the skyrocket was the asteroid rocket, which contained as an additional payload element, a parachute. At the peak of the rocket's trajectory, the head would be released, the parachute ejected and deployed, and the rocket's slow descent made with bright-colored flares burning all the while. Triplet asteroid rockets, with three parachutes, are described in Thomas Kentish's *The Complete Art of Firework-Making.*

Artificers from the sixteenth century on developed "ground" rockets, that is, rockets that did not fly into the air but rather activated some device or display on or near the ground. One version was the line rocket, or *courantin,* that propelled figures along tautly stretched cords. Another was the wheel rocket which, when fitted to the circumference of a wheel, would cause it to spin. A variation of the ground rocket was the water rocket, which either would skim along the surface of the water or,

". . . and here may you note that upon the *rockets* may be placed fierie *Dragons Combatants,* or such like to meete one another, having lights placed in the Concavity of their bodies, which will give great grace to the action." Early seventeenth-century print (see page 16).

occasionally, dive beneath, emerging in successive arcs. Robert Jones, a lieutenant in the Royal Regiment of Artillery, described such rockets in 1765 in his *New Treatise on Artificial Fireworks.*

> Water rockets may be made from four ounces, to two pounds, but if larger they are too heavy, so that it will be difficult to make them keep above water, without a cork float, which must be tied to the neck of the case, but the rockets will not dive so well with, as without floats.
>
> Cases for water rockets, are made in the same manner and proportion as sky rockets, only a little thicker of paper; when you fill these rockets which are drove solid, put in first, one ladle full of slow fire, then two of the proper charge, and on that one or two ladles of sinking charge, then the proper charge, then the sinking charge again, and so on, till you have

> filled the case within three diameters. . . . When you fire these rockets, throw in six, or eight at a time; but if you would have them all sink, or swim, at the same time, you must drive them with an equal quantity of composition, and fire them all together.

A variety of compound fireworks was developed over the years, probably the best known being the girandole. The word was taken, as so many pyrotechnic terms were, from the French *girandelle,* which in turn came from the Greek word for gyrate or revolve. Some early English and French artificers used the word to mean revolving fire wheels, but eventually girandole came to refer only to simultaneous flights of many rockets, or "fountain" of rockets, to take Cutbush's term. Thomas Kentish tells how to fire a girandole of a hundred rockets at once.

> Suppose a cubical tea-chest. In the top, bore ten rows of holes, ten in a row, with a centre-bit; the same in the bottom, in such a way that the bottom holes fall perpendicularly under the top holes. Fasten the box upon four legs, one at each corner. Sift from a pepper-box a layer of meal powder over the top; put in the rockets, with their primed mouths, naked, to rest on the sifted meal. It is evident that, upon conveying fire to the meal by a leader, the flash will ignite the whole of the rockets at once. Of course it ought not to be a tea-chest, but a box constructed on purpose, with a penthouse lid, to fall over and protect the rockets till desired to be fired.

Not until the nineteenth century did the military version of the European rocket come into the forefront. Curiously, it did not receive its direct inspiration from earlier British, French, or other experience in Europe, but rather from events taking place in far-off India (Chapter 3). There, the rocket troops of Haidar Ali and later Tippoo Sahib of Mysore had unleashed large numbers of rockets against British forces, causing a considerable annoyance. Before long, these rockets—which were made of iron casing lashed to bamboo canes and weighed no more than a couple of pounds—came to the attention of the English artillery expert (later, Sir) William Congreve (1772–

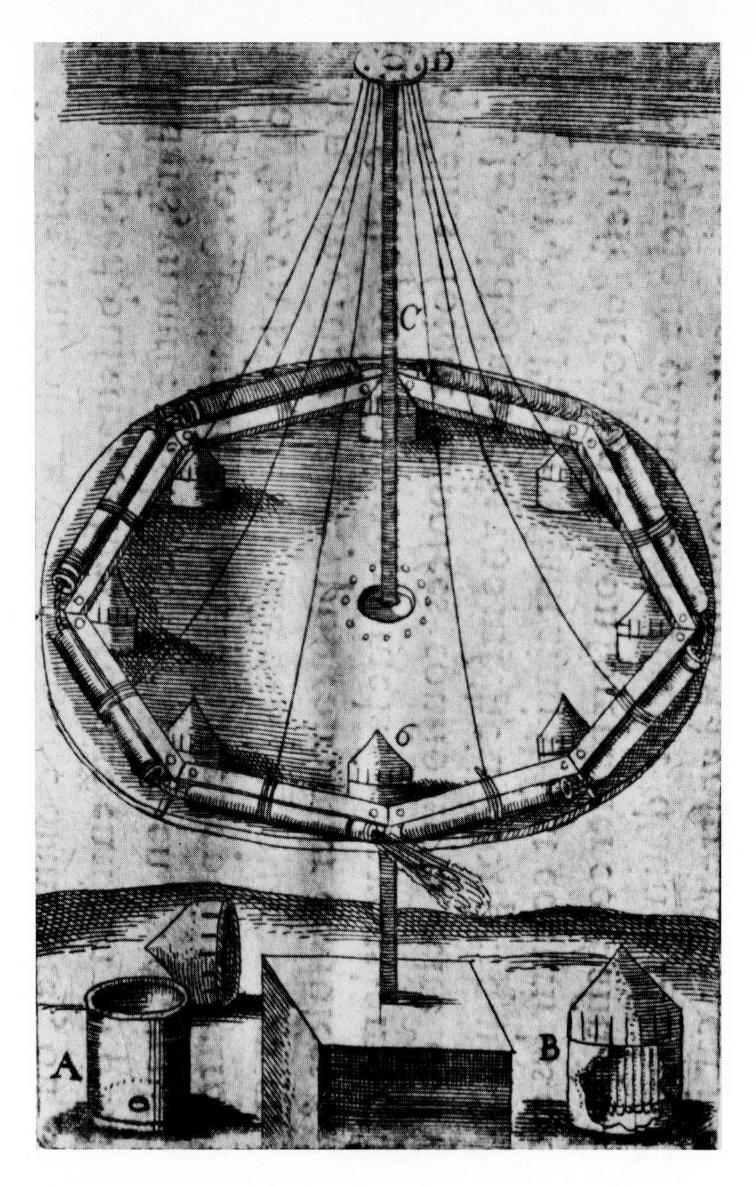

The *girandelle* (or girandole, as it became known in English), a revolving rocket fireworks device popular to this day. Sometimes the term "fire wheel" was used to describe it, while "gironel" and "girande" are also found. The one depicted here is early seventeenth century. Each time a lateral rocket ignites, it in turn ignites one of the boxes containing "serpent" fireworks or squibs shown in detail at *A* (empty) and *B* (filled). The whole device turns on axle *C* attached to screw plate *D*.

1828), who used them as the basis for his subsequent and successful development of an impressive series of land and naval weapons. He fitted his missiles with either case shot or shells if they were to be used for field operations or with an incendiary mixture if their purpose was bombardment and conflagration. The latter type, called a "carcass rocket," was loaded with a hard-packed incendiary composition and first saw action in 1805. All Congreve rockets had metallic cases and were stick-guided. Depending on the particular weapon, the sticks would be longer or shorter and would normally be both stored and carried in two or three easily connectable pieces.

Congreve rocket ammunition was divided into light, medium, and heavy categories, the first being 6- to 18-pounders and employed principally for field service, the second ranging from 24- to 42-pounders, and the third any rocket larger than a 42-pounder. (We recall that rockets were often designated by the weight of a lead ball that could fit into the powder-loading mold.) The 42-pounder plus rockets were also often designated by their caliber; an 8-inch "exploding" rocket might weigh up to 300 pounds, of which 50 pounds represented warhead. The range for such a weapon varied between 2,000 and 2,500 yards. At the other end of the scale was the 6-pounder, the smallest and lightest carried in regular inventory. It could deliver 3 pounds of shot or shell.

The force of Congreve's rockets was considerable. A 32-pounder, for example, could penetrate the walls of buildings or up to nine feet of common earth at a range of 3,000 yards. The 12-pounder case-shot rocket was especially popular because it was easy to transport and fire and could reach nearly twice the range of artillery of the day. Cutbush pointed out that such rockets enjoyed a number of other advantages:

> We may remark here that the projectile force of the rocket is well calculated for the conveyance of case shot to great distances; because, as it proceeds, its velocity is accelerated instead of being retarded, as happens with any other projectiles; while the average velocity of the shell is greater than that of the rocket only in the ratio of 9 to 8. Independent of this, the case shot conveyed by the rocket admits of any desired in-

The mode of using Congreve rockets in bombardment, showing bombarding frame and ammunition being carried by troops. Basically, the apparatus is a light pole about 10 feet long with a 24-pounder or 32-pounder rocket tube fixed by staples, and two light legs that support it in triangular fashion when deployed. In position for firing and ready for action, the command is given "Elevate to thirty-five degrees!" (or whatever angle is desired). Then the word "Point!" is given, which means that the plumb line hanging from the vertex of the triangle must be checked to determine whether or not the frame is upright. At "Load!" a rocket is inserted in the lower end of the tube, the end of the stick resting in the ground. The trooper then takes the trigger line and steps back ten to twelve paces and awaits the order "Fire!" A 32-pounder could typically reach up to 3,000 yards distance.

crease of velocity in its range by the bursting of powder, which cannot be obtained in any other description of case.

Rockets and their launchers were moved from location to location by soldiers, by horses, or by horse-drawn ammunition carts specially designed for the purpose. If soldiers alone were involved, a hundred men would handle three hundred rounds and ten launching frames. A good rocket-firing team could set off four rounds a minute. Alternatively, four horses could carry seventy-two rounds and four frames from which an experienced crew could make up to sixteen launches a minute. Eighteen, twenty-four, and thirty-two pounders were often hauled in carts, which could be readily converted to serve as launching stands. Two rockets would take off at a time from double-iron plate troughs that could be moved from horizontal point-blank position upwards to a 45-degree elevation.

Rockets became popular with the British navy as well as the army primarily because they were light, mobile, and produced no bothersome recoil. Moreover, they were inexpensive: a 32-pounder carcass rocket cost £1/1s/11d in the 1820s. This rocket was larger than those normally used by the navy, the more common ones being 8-, 12-, and 18-pounders that could be fired from such vessels as four-oar gigs and eighteen-oar launches. When launched against land targets, rockets would follow high-angle trajectories; but, if an attack against shipping was involved, they would occasionally be fired at low angles so as to ricochet across the water and skip into enemy portholes. In explaining his plans for a rocket attack on the harbor at Boulogne, France, Congreve wrote to His Royal Highness the Prince of Wales, on February 20, 1806:

> In this project, it was proposed to equip ten men of war launches: so that each launch, rowing eighteen oars, might, in a volley, discharge fifty 8-pounder Rockets, each Rocket containing as much carcass composition as an eight-inch spherical carcass. These launches were to be towed across by the gun brigs of the squadron— When the attack was to be made, they were to row in, by night, within 2,000 yards of the bason [basin], to certain stations, marked in the chart, led in by some

The use of rockets in British fire ships in early nineteenth century. Congreve, in several publications, proposed stowing tier above tier of rockets in the fire ship. When the time approached for engaging the enemy, the rockets were placed in racks, at different angles and in all directions. When the fire ship drifted "in amongst the enemies' ships" a tremendous volley would be released. Congreve felt that the enemy would not attempt to tow the fire ship clear, "as it is impossible that any boat could venture to approach a vessel so equipped, and pouring forth shell and carcass Rockets, in all directions, and at all angles." The illustration at bottom, center, is a close up of a fire ship in action; the illustration at left shows the rocket armament in the *Galgo* "defence ship"; the illustration at right is a close up of the bulkhead.

experienced officer. From these points the Rockets would range well into the bason, and the launches not be nearer than 1,000 or 1,200 yards to any battery. The Rockets were laid in frames, fixed at 55° of elevation, between the masts of the launches, so that they could be fired by a leader terminating in a gun lock, 50 in a volley—the ten boats, therefore, would discharge 500 in one flight, which would, by one single and momentary operation, convey as much carcass matter as could be thrown in four hours by ten 8-inch mortars; so that the boats need scarcely be five minutes exposed to any fire, as they would have only to row rapidly down to the desired points, and to retreat as rapidly after discharging the Rockets.

Congreve knew as well as his ordnance critics that the conventional stick rocket could only be improved to a point, that it was doomed to extinction as a weapon of war unless its accuracy could be improved. One modification was to diminish the length of the guidance stick, which not only made the missile more portable but subjected it to less deflection by wind gusts. Somewhat later, Congreve changed the stick from its awkward lateral position and attached it centrally. This increased flight efficiency and facilitated the launch process, for it meant that tubular launchers could easily be employed. But much remained to be done.

"Like most inventors, Sir William Congreve was sanguine," wrote John Scoffern in 1852 in his *Projectile Weapons of War and Explosive Compounds,* and, having made several improvements, Congreve "imagined that his war rocket would completely alter the practice of artillery; and would be all but universally applicable, whether for the purpose of battering fortifications or slaughtering men." Scoffern recognized that rockets had some distinct advantages. Compared with an artillery shell, they were not size-limited—not requiring heavy ordnance to fire them, rockets could be made any size and were extremely portable. Also, they did not pose the recoil problems that artillery did—and still does. Rockets like artillery, could be mass-fired almost simultaneously. Not to be underestimated were the physical and psychological effects wrought by the rocket. "Cannon-balls and shells," observed Scoffern,

The use of rockets on boats in the early nineteenth century. Two launches operating from a British man-of-war are illustrated here. The same frame for launching Congreve rockets on land is being employed for sea duty, except that its support legs have been removed and the vessel's foremast used instead. The launch tube can be raised or lowered with halyards. Firing is about to take place in the lead boat, the rocket being ignited through a trigger line leading aft. In the second boat, the loading of a rocket can be seen, with the frame having been lowered to a convenient height. In order to ascertain that rocket exhaust sparks do not ignite the sail, the latter is kept wet at all times. This was by no means indispensable, wrote Congreve, "as I myself discharged some hundred Rockets from these boats, nay, even from a six-oared cutter, without it [e.g., without wetting the sail]." The rocket ammunition is stored in the launch's stern sheets, covered with either tarpaulins or tanned hides. If a six-oared cutter is used, there is no room for storage, so an attendant boat is necessary. Congreve never ceased to point out a major advantage of the rocket for sea warfare, to wit, "its property of being projected without reaction upon the point of discharge."

> rush on their objects unseen, or faintly visible; not so the rocket, which carries with it a long fiery tail, burning everything in its course, exploding ammunition wagons, mowing down troops, and producing amongst cavalry the most inextricable confusion. No horse, however well disciplined in regard to other artillery, will stand the hissing of a rocket; not even the horses of a rocket troop.

Yet doubts persisted. First a shifting center of gravity had to be taken into account, Scoffern noted:

> If the rocket could be endowed with the precision of which cannon are susceptible, it would indeed accomplish all that Congreve hoped; but such precision has not yet been attained, and this for obvious reasons. In the first place, the theoretical determination of a rocket's flight is a much more difficult affair than the determination of the flight of a shot or shell; because in the former case we have a constantly diminishing weight (occasioned by the burning of the composition), producing a continued variation of the centre of gravity; whereas in the latter case the weight never varies, and its centre of gravity never shifts.

Then there was the much more serious difficulty of guiding the rocket accurately through the atmosphere. Scoffern used the term "deflecting agency of currents of air," explaining that

> If such currents are capable of producing a manifest effect on the flight of shot and shells, bodies either homogeneous, or nearly so, how much greater must be this influence on a missile formed like a rocket, which departs as far as possible from the spherical form, in regard to freedom from the interfering causes of atmospheric currents.

He might also have added that the guiding sticks of the day were never perfectly straight, nor were they all centrally attached, leading to inevitable course deflections. Moreover, even if an individual stick were absolutely straight and properly mounted along the line of the rocket's axis, it was still flexible and susceptible to deviating the flight course. Finally, gusts of wind upon the stick made it act as a weathervane, further disturbing the desired trajectory.

The only obvious solution to the accuracy shortcoming was to do away entirely with the cumbersome stick. But how, then, would stable flight be achieved? Theoreticians and experimentalists alike knew the answer: give to the rocket a rifled or rotary motion. Since the rocket is not fired from a gun, the rifled motion achieved by a gun barrel was out of the question; this meant that the rotary motion would have to be created by the rocket itself. Attempts were made to achieve this by attaching wings and vanes to the case and by cutting spiral grooves into the case, but with little success.

Around 1839 William Hale, a self-educated naval designer and ordnance expert living in Greenwich, directed his attention toward rockets. Within a few years he devised a scheme whereby he could give rotary motion to a rocket by directing part of the exhaust flame through slanted, peripheral apertures. In 1844 he was awarded a patent for his "stickless" or "rotary" rocket, and in 1863 Hale published his *A Treatise on the Comparative Merits of a Rifle Gun and Rotary Rocket.* During the intervening years he improved launchers for both land and naval rockets.

When he went to work on the rotary rocket, Hale was not only faced with the problems of a shifting center of gravity and atmospheric instability, but also with the fact that rockets of the period could not be manufactured with precision. As he put it, "one part is more obtuse than another, and consequently the direct action of the atmosphere is greater on one side than another, and has a tendency to turn the rocket out of its course." He had also to cope with the fact that rocket apertures were never machined perfectly, and hence escaping exhaust gases, "not being wholly in the direction of the rocket's axis," necessarily had to produce some deflection of the rocket from a true flight path.

By spinning a rocket, he predicted—and later proved in tests—conditions would be improved measurably, for then the "whole cylinder must move 'en masse.'" He added:

> While the resistance of the air affects one side more than the other, the cylinder will be changing its position until the

> quantities acted upon by the resisting medium around the centre of gravity are equal, which will be when it is in a vertical position, and in this state it will continue to fall till it reaches the earth.

Moreover, he observed that

> if the rocket have a rapid rotary motion, the centre of gravity and the axis of motion will of necessity coincide, and all irregularities of surface as to angular position in the head of the rocket, and all imperfections in the vent, will be compensated by the rotary action. The use of the stick being dispensed with in consequence of rotation, it is almost unnecessary to remark that the wind cannot act upon a rotary rocket as upon the Congreve, and that the imperfections appertaining to the use of the stick in the rotary rocket will have no existence.

Hale's success was widely recognized by military ordnance experts in Europe and the United States and resulted in many letters of commendation. One, written on May 27, 1850, by Robert J. St. Aubyn, a lieutenant in the Royal Navy stationed at the Coast Guard Station at Shoeburyness, took note of the long range attained:

> Sir,
>
> I have the pleasure to inform you that one of the large rockets fired on the 25th was picked up about three-quarters of a mile in advance of the last range picket, or in distance about 100 yards short of three miles off the point from which it was fired; the other, which was the one with small brass screws in the heads, fell about 400 yards short of the above distance.

War Office trials conducted in 1856 showed deflection from target of only ten yards for 12-pounder rockets traveling over a 4,600 yard range, eight yards for the 24-pounder at 4,200 yards, and nine yards for the 100-pounder at 3,300 yards. Ordnance Major William H. Bell stated that "the Secretary of War appeared to be highly pleased with the result." The Americans were equally impressed: a joint board of army and navy officers concluded as early as 1846 that "the effect of the rockets, with

Lifesaving rocket launcher in position along the shore. When igniting the rocket, the operator placed himself to one side.

regard to range, force, and accuracy, is at least equal, and probably superior, to that of the ordinary Congreve rocket of the same size," and, "The fact of this rocket being without a stick gives it an incontestible superiority over the Congreve rocket. . . ." The board then recommended that "an arrangement be made with the proprietor [Hale] for the purchase of the full instructions requisite for making these rockets."

The advances of Congreve, Hale, and other rocket technologies in Britain and similar efforts in such countries as France, Russia, Germany, Italy, and Austria, combined to give rise to new applications of rocketry—new in the sense that they were not oriented toward the military nor toward fireworks. Of these, two stood out: the lifesaving, marine, or succouring rocket and the whaling rocket. The first was designed to carry a cord or rope (to permit rescue operations) from the shore to shipwrecked vessels, or from such vessels to the shore, while the latter was used to propel a modified harpoon into aquatic quarry.

Among the pioneers of lifesaving rockets were Henry Trengrouse, William Congreve, John Dennett, Alexander Carte, Edward Boxer, and William Schermuly. Mitchell R. Sharpe, in his comprehensive essay "Development of the Lifesaving Rocket," points out that in the British Isles alone at least 15,000 lives were saved by such rockets between 1871 and 1962. During one storm, on November 10, 1810, sixty-five ships and small boats were wrecked in the North Sea, "most of them less than three hundred feet from the coast." Despite the employment of large quantities of military rockets during the nineteenth and twentieth centuries, Sharpe speculates that "Paradoxically, more lives may have been saved by the rocket in [the nineteenth] century and the succeeding one than were lost to it."

John Dennett of Newport, on the Isle of Wight, was one of many inventors to work with, develop, and/or improve the lifesaving rocket. Preferring 12- and 18-pounders, he explained in 1832 that they are fired by

> means of an excellent brass-framed lock, (either *flint* or *percussion,* at the option of purchasers) from a light portable stand, with a graduated arc, and horizontal movement, for elevating and pointing; which for the shore (or land service) is supported on legs, and for sea-service fixes in the quarter-rails, and other convenient parts of the ship . . . The larger Rockets are contrived to fix themselves to the rigging, or other parts of the wreck, to be used in case the crew . . . be so much exhausted, or benumbed with cold, as to be unable to assist themselves by making the line fast; under such circumstances, a boat may be hauled off through the surf by means of the line.

Rockets to be fired from ships onto the shore were constructed "for penetrating, and holding fast in the ground, in situations where no persons are on the beach; so that if a boat can be got overboard, it may be hauled on shore by the line; or otherwise, a man may secure his landing through the surf, and perform the subsequent operations for saving the rest of the crew." Dennett also visualized his rockets being employed to hurl to shore mail bags and dispatches from ships in danger of breaking up.

One of the greatest triumphs of a Dennett lifesaving rocket occurred on October 8, 1832, when the 430-ton *Bainbridge* was wrecked on Atherfield Rocks, Isle of Wight, in what was described as "a dreadful Gale." The grateful captain, William Miller of Halifax, and the mate, Joseph Irvin, prepared a certificate of appreciation in which they stated that one of Dennett's rockets

> at once carried the line to its destination in the most complete manner, although the position of the ship was most unfavourable for such a manoeuvre, as she lay *end on*, with her stern towards the shore. The communication thus established, a strong rope was hauled on board, and a boat drawn through the surf, by which the crew, nineteen in number, were in two trips safely landed upon the beach . . . we heartily wish, and we give our most hearty thanks, (as is most due) to Mr. Dennett, for his exertions in the cause of humanity.

The feature that most attracted whalers to the rocket was the lack of recoil and the velocity at which an attached harpoon could reach moving, partly or wholly submerged targets. Although the first efforts to make a whaling rocket are credited to the seventeenth-century Dutch pyrotechnician Abraham Speeck of Amsterdam, it was not until the nineteenth century that such rockets gained popularity.

One instance of the success of a whaling-rocket operation appeared in the Hull (England) *Advertiser* on October 12, 1821. The ship *Fame* under Captain Scoresby "brought home nine fish [whales], in the capture of the whole of which the rockets were successfully employed." The article went on to say:

> After being struck by the rocket, the largest whale became an easy prey to its pursuers. In one case instant death was produced by a single rocket, and in all cases the speed was much diminished, and its power of sinking limited to three or four fathoms. One of the largest finners, of 100 feet in length, a species of fish seldom attacked by the ordinary means, and of capture of which there is scarcely an instance on record, in the Northern seas, was immediately tamed by a discharge of rockets, so that the boats overtook and surrounded it with ease. Six out of the nine fish died in less than fifteen minutes and five out of the number took out no line at all . . . The peculiar value and importance of the rocket in the fisheries, is, that by means of it, all the destructive effects of a six-, or even a

twelve-pounder piece of artillery, both as to penetration, explosion, force and internal fire, calculated to accelerate the death of an animal, may be given with an apparatus not heavier than a musket, and without any shock or reaction on the boat.

The Americans as well as the English found merit in the unfortunate application of rockets to the slaughter of whales. Captain Thomas Welcome Roys of New York, who later became the first to introduce whaling rockets to California, had witnessed the power of the rocket during the siege of Oporto in the 1832–33 Portuguese civil war. Back in New York, he joined with a well-known pyrotechnician, Gustavus Adolphus Lilliendahl—a man of both whaling *and* fireworks interests—and together they developed and patented a whaling rocket. The new device was tested with moderate success during the 1865–66 Arctic whaling season. Moving to California, they continued their work there until the mid-1870s.

Later in the same decade, John Nelson Fletcher and Robert L. Suits established the Fletcher, Suits & Company, California's first full-fledged whaling rocket enterprise. According to Frank H. Winter and Sharpe in an article on whaling rockets, the two men found the Roys-Lilliendahl rocket harpoons too light and "charged with too little powder." They decided, therefore, to produce a 6½-foot-long rocket bomb lance that weighed 32 pounds and yet was so simple that it could be operated by "a cabin boy." Soon, Fletcher and Suits began to boast that their rocket "could fasten to a whale at 30 fathoms [180 feet], which was a considerably greater distance than could be reached by hand harpoons or even gun lances." During 1878, they killed 35 humpback, sulphur bottom and fin-back whales with their rockets, which Fletcher and Suits described in the following terms:

Our apparatus consists of a gunmetal cylinder, filled with a peculiar composition made only by ourselves, to which is attached, in front, a bomb with a barbed point; inside the bomb is an explosive charge and a chain toggle, which is released by the bursting of the shell on entering the whale; an iron shaft is attached to the rear of the rocket, through which the

A rocket brigade at work saving a passenger from a foundering ship in the late autumn of 1898.

> whale line is spliced. There is absolutely no recoil . . . the hinged flange is thrown up by the rocket passing out, protecting the face from injury.

Despite claims of success, by the late 1880s whaling rockets began to disappear. Whales had for some time been overhunted and were getting harder and harder to find. This drove up costs per creature killed. Then, too, petroleum and kerosene products were increasingly being substituted for whale oil, whose price dropped dramatically. Also, rocket harpoons were expensive to make and never did reach a satisfactory state of reliability. Finally, there was a more subtle drawback pointed out by Winter and Sharpe: "Veteran mariners were apt to regard the newfangled whaling rocket or anything else so radically innovated with the gravest skepticisms and pre-formed prejudices."

Whaling rockets and all other forms of rockets suffered a serious decline as the nineteenth century ended. Firework displays became less fashionable; military rockets gave way to rifled, breech-loaded, rapid-firing artillery; larger ships, better navigation, and telegraphic communication reduced to an extent the need for lifesaving devices of all types. Almost no interest was exhibited in rockets during the opening years of the twentieth century, but with the approach of World War I attention focused on improving signal and illumination devices of several types.

Building on their eighteenth-century fireworks experience and on nineteenth-century military rocketry, the French sought a solution to the vital problem of lighting up battlefields at night. Well before the outbreak of the great conflict of 1914–18, pyrotechnicians at Bourges had developed a rocket that could fly up to 1,000-foot altitudes, then release a parachute that lowered a tube of burning composition. Depending on the parachute, the weight of the composition, and the altitude attained by the rocket, the device would provide a cone-shaped, down-directed light for from forty-five seconds to several minutes duration. During the war, such rockets lit up enemy trenches and no-man's land, enabling machine-gunners and riflemen to pick out their targets at night.

Soldiers fired similar rockets throughout the war to communicate between sectors, between front lines and advanced posts,

Naval Lieutenant Y. P. G. Le Prieur at Verdun, during World War I, with a "Bébé" Nieuport pursuit plane loaded with his rockets in the background. (Courtesy Musée de l'Air, Paris)

and between ground forces and observers in airplanes and balloons. Signal-rocket communications could only be cut off by the enemy if the site from which they were being launched was captured. Since signal rockets were extremely portable, this was of little concern. One German model consisted of a sheet-iron tube with a brazed and riveted open-hole brass plate at the lower end. Most of the tube was filled with propellant, above which was the igniting powder. The canister head was lined with brown paper and filled with illuminating stars. On occasion, the British and Italians modified their signal rockets so they could be used as message carriers. Messages, including maps, would be placed in metal tubes attached to the lower end of a guiding stick.

That sticks were still around some seventy years after Hale's invention of the rotary rocket may sound strange, yet it does demonstrate how the rocket had retrogressed from disuse. Indeed, much mid-nineteenth-century rocket technology was either unknown or not available to twentieth-century ordnance experts hurriedly preparing for war. One only has to read the opening paragraph of the entry "Fusée de guerre" (war rocket) in the 1898 edition of *Dictionnaire Militaire* to comprehend what had happened before World War I.

> After having enjoyed, for more than half a century, a vogue that often touched of exaggeration, the war rocket today is almost completely abandoned. It is nevertheless necessary to treat the subject not only for historical reasons but in the hope —not yet abandoned by everyone—that in the future the device will be resurrected because of its simplicity and ease of deployment.

At the beginning of World War I, Congreve stick-guidance-type rocket technology was resurrected in France, mainly because of this simplicity and ease of deployment. Toward the middle of 1917, General Ferdinand Foch ordered the French army's Section Technique de l'Aviation to study the feasibility of small-caliber rockets for combating German observation balloons called "Drachen." The section immediately called on the talents of naval Lieutenant Y. P. G. Le Prieur, who had recently been working with rockets manufactured by the French fireworks manufacturing firm Établissements Ruggieri. (He was experimenting with rockets fired into clouds so that hail storms would break over uninhabited areas rather than over farmland and towns.) Le Prieur quickly adapted stick-guided, powder rockets for firing through steel tubes mounted in vertical rows of four or five on the wing struts of pursuit biplanes like the Nieuport. During the great Allied offensive of 1917, the rockets proved successful, as pilots let loose their barrages in accordance with set procedures, per the pilot instruction sheet:

> This fire [against captive balloons] is effected at a distance of from 100 to 150 [meters], about, while diving at an approximate angle of 45° at least. The steeper the dive, the straighter the trajectory and the more effective the fire. The attack should always be made in the *direction of the length of the balloon* and *against* the wind. If there is no wind, sight should be taken

directly on the center of the balloon, but if there is an appreciable amount of wind, correction in sight should be made as follows: The mean speed of the rocket being about 100 meters a second and as the discharge should be made at a distance of about 100 meters from the balloon, the wind will give to the rocket a relative recoil equal to the speed of the wind in meters a second. Sight should then be taken at a point in front of the center at a distance equal to this recoil. . . .
Extremely Important Remark:
The departure of the rockets does not follow immediately the touch of the electrical button [in the cockpit], and the delay varies slightly from one rocket to another. Therefore it is absolutely necessary to continue to hold the sight and the descent until the discharge of the last rocket (about one second). If one redresses or turns too quickly, the last rockets will go in different directions and give a dispersion which is altogether inadmissible.

Before closing this chapter, we must describe the work of Wilhelm Teodore Unge (1845–1915). It is difficult to know just where this Swedish military engineer belongs in the history of rocketry, for he labored with little official encouragement outside the main stream of events connecting Congreve and Hale with World War I, and the results of his impressive developments were not applied until long after his death. Also, he had exhibited no interest in rockets until the 1880s, a time when rocket corps all over Europe were being disbanded and manufacturing facilities closed down. (In Austria, for example, the rocket forces established in 1815 by Major Vincenz Augustin were abolished in 1867.)

With the financial support of Alfred Nobel (1838–1896) millionaire inventor of dynamite, Unge established the Mars Company in the late 1880s, to improve the performance and accuracy of rockets. He first tried to increase range and accuracy by firing a rocket from a cannon (to provide initial, or first-stage, velocity) and then igniting the propellant charge (to give second-stage thrust). Though the technique was sound, and would be proven so in practice in the 1960s, when Unge made his tests in the early 1890s he met with failure. The problem: the high

Solid-propellant rocket developed in the late nineteenth century by Wilhelm Teodore Unge of Sweden. Rocket is being launched, spin-stabilized by a rotating launcher that is pneumatically driven.

initial acceleration damaged the rocket's powder charge ignition system, and the second stage therefore never fired.

This cannon approach proving unfruitful, Unge—with the continuing encouragement of Nobel—turned toward improving the propellant powder. Curiously, no work had been accomplished during the nineteenth century toward developing more powerful powders, which had remained essentially as they had been in the seventeenth and eighteenth centuries. Unge took maximum advantage of Nobel's work in double-base ballistite smokeless powders, which consisted of nitrocellulose and nitroglycerin, and soon came up with a controlled burning mixture that produced a higher exhaust velocity than traditional powders. To Nobel's double-base powder, Unge added a stabilizer (which served to retard chemical decomposition during storage) and a plasticizer (which increased the propellant's plasticity or workability).

A man of the 6th York and Lancaster Regiment fixing SOS rockets in a front line trench in France. Cambrin, February 6, 1918. (Courtesy Imperial War Museum, London)

Later, in co-operation with the Skanska Bomullskrutfabriks company, he developed a binder that gave greater mechanical strength to the propellant grain. Unge's first ballistite-powered rocket flight took place on September 12, 1896.

As if he had not accomplished enough in making a pioneering improvement in rocket-propellant chemistry and technology—something his predecessors in other parts of Europe had barely considered—by the end of the nineteenth century Unge was attempting to outdo Hale. Rather than spin the rocket itself, as the Englishman had done, Unge rotated the launcher. His approach was sound, and he soon received three patents covering three different approaches. By 1905, launcher-rotated rockets were reaching five-mile ranges with accuracies that competed with the rifled artillery of the day.

Unge saw many applications for his rockets. He suggested surface-to-air versions to be used to knock down enemy balloons. Ship-to-ship types would prove invaluable during naval engagements, he said, while ship-to-shore missiles would assist conventional cannon in softening up fixed positions during a general naval bombardment.

The Swedish military turned a deaf ear.

In fact, no one listened at all except the mammoth Krupp enterprise in Germany. In 1908 Krupp interests bought out Unge, patents and all. Yet the munitions company did not employ Unge technology before or during World War I. A major thrust forward in rocketry had stalled.

After its brief resurgence during the 1914–18 war, the rocket again went into a decline as the great powers disbanded their military forces and the research and development establishments behind them. When the rocket reappeared, it was under conditions and motivations far changed from those of the past.

Sir William Congreve. (Painting by James Lonsdale)

6

THE AGE OF CONGREVE

Here we end our progression into the twentieth century, regress about a hundred years, and focus on the opening of the "Age of Congreve." It was not until the early 1800s that the rocket finally became an important element of war, first in the British forces and later in those of most of the major European powers, including Russia. Before the nineteenth century was over, hundreds of thousands of rockets had been manufactured and fired under an astonishing variety of circumstances.

They were fired, for example, during the bombardment of Copenhagen in 1807, at Waterloo in 1815, in the Crimean War in mid-century, and on many other occasions. Thousands were unleashed against poorly armed adversaries during the rise of colonialism. The British set them against American forces during the War of 1812, in China during the Opium War in the 1840s, in Africa during a series of engagements during the latter half of the nineteenth century, and even in remote Peru during its independence movement in the early 1820s. The Russians, French, Austrians, Italians, Danes, Swedes, Spanish, and others launched rockets in both foreign and civil conflicts. The rocket became a familiar sight from one end of the earth to the other, and its fiery red glare even inspired a line in the American national anthem.

Rockets inspired their promoters, as well. Napoleon's Marshal Auguste Marmont, Duke of Ragusa, exclaimed: "I repeat it, Congreve rockets must produce a revolution in the art of war; and they will assure success and glory to the genius who shall have been the first to comprehend their importance, and to develop all the advantage which may spring from their use."

William Hale, feeling he had done exactly that, agreed: "The rotary rocket will henceforth rank as second to none of all the destructive engines that war unhappily brings into operation."

The technological climate created by the Industrial Revolution was responsible for the impressive ascendancy of the war rocket in Great Britain where the revolution began. Progress was so rapid that in 1807 the English fleet was able to fire clouds of incendiary rockets on Copenhagen, setting fire to much of the city. Other nations were impressed and quickly set about introducing rockets into their own military services.

Despite early successes, the war rocket had its detractors as well as its champions. Many politicians deprecated the new weapon because of the commotion and nuisance it created, and an occasional newspaper editorial decried its devilish qualities. The Pope, however, did not outlaw the rocket as an earlier pontiff had that "ultimate" weapon of his time, the crossbow. Sober-minded artillerymen and ordnance officers recognized that even though the rocket had improved more in the first decade of the nineteenth century than in all the years of its previous existence, severe performance limitations remained. Uneven progress in physics, chemistry, and metallurgy resulted, for example, in much improved rocket cases but only marginally better propellants. Not until the mid-twentieth century would the problem of accuracy finally be solved.

While several European nations could each claim a rocket pioneer with varying degrees of legitimacy, three men emerged during the nineteenth century who made major and basic contributions to the design and development of the weapon: William Congreve, William Hale, and Wilhelm Unge. As in most military weapons research, secrecy surrounded their and other rocket projects to the extent that some nineteenth-century developments did not survive to the present day. At the same time, other rocket experts openly sold their talents and services abroad, revealing to foreign governments the very secrets their governments were trying to withhold. One Westermaijer, a mechanic at an Austrian rocket manufacturing plant near Wiener Neustadt, went off to Warsaw where he helped the Poles develop military

Rocket practice in the marshes near Woolwich Arsenal, 1845. Contemporary print.

missiles; later he sold his services to the Prussians, and still later he made contact with the Russians and the Swedes.

In a similar manner, William Bedford, who had worked for Congreve in Britain at the Woolwich Arsenal, moved across the channel to France where he remained for twenty years directing rocket research and development. Other qualified "free-lance" experts offered their services wherever they could find buyers, particularly to countries without native expertise. Thomas Williamson, an Englishman who claimed to be the true inventor of the Congreve rocket, offered his talents to United States President James Madison in 1816, though French military authorities were avidly bidding for them.

There were, of course, countries unable or unwilling to develop their own rockets, but there was always a ready supplier. Several newly emerging nations in South America, such as Colombia,

Mexico, Uruguay, and Argentina, had access to rockets, many of which were furnished by Congreve himself from his private rocket factory at Bow, near London. While Congreve served as superintendent of the Royal Arsenal at Woolwich developing war rockets for the British army and navy, his private plant was producing them for sale in other countries. But he also sold them to the British government for use in campaigns against the Indians in Canada in 1816. Congreve's business manager at the Bow plant approached the Dutch and demonstrated the rockets to them in 1830.

William Hale patented his invention (the spin-stabilized rocket) in the United States and then sold the process to the U. S. Army Ordnance Department in 1846. His ideas were not adopted in his native Britain until the 1860s. Hale's son traveled to South America, selling the rocket there as the local wars of liberation began.

Aside from Congreve and Hale, very few gave consideration to the tactical employment of the weapon as it then existed. Artillerymen were upset over the rocket's lack of accuracy when compared to cannon shells, overlooking the point stressed by Congreve that "the very essence and spirit of the Rocket System is the facility of firing a great number of rounds in a short time, or even instantaneously, with small means." Much time and many resources went into efforts to improve the range and correct lateral dispersion (errors to the left or right of desired line of flight) of the rocket, to the detriment of its use as a very effective weapon for delivering massed firepower on an area target. Practically all military commentators of the time, however, stressed its advantages as a pyschological weapon and as an anticavalry arm.

In 1804 Congreve first realized that rockets would be especially useful as weapons because "the projectile force . . . is exerted without reaction upon the point from which it is discharged." (The bothersome recoil of conventional artillery was thus not present.) He foresaw rockets being fired from ships as well as by land troops and began working hard on their development. So rapidly did he progress that by late summer of the same year the first contingent of Royal Marine artillerymen arrived at Woolwich Arsenal to receive training in the theory, fabrication and firing of his rockets.

Enjoying the fullest support of his father, who was a major general, of the Board of Ordnance, and of the Inspector of Artillery, young Congreve continued improving his weapons until by August 1805 he had convinced Lord Castlereagh, the Minister of War, that they could be used against the French fleet harbored at Boulogne. This fleet, the Admiralty suspected, was the nucleus of a force Napoleon was readying for the invasion of England.

Although Prime Minister William Pitt, Lord Castlereagh, and others were solidly behind the rocket bombardment scheme, Lord Keith and the Admiralty were not. Naval officers spoke contemptuously of "Mr. Congreve's squibs" and planned instead to strike a sudden blow at Admiral Villeneuve's fleet at Cadiz. However, to the disgust of the Admiralty, Castlereagh selected Commodore Sir Sidney Smith to lead a rocket attack on the French fleet at Boulogne. Smith, a flamboyant character not adverse to publicity, saw the merits of the rocket and wholeheartedly endorsed the attack. Finally, King George himself approved the project, and so, grumbling, the Admiralty had no choice but to accept it.

The plan called for special marker boats to move to a designated position and anchor; rocket boats were to follow in line behind, circle around the marker boats, fire their rockets, and then withdraw to the protection given by mortar and bomb ships. In mid-October Congreve assured Sir Sidney that everything was on schedule.

The Royal Marine artillerymen training off Dover had been given one supply tender for every rocket launching boat. Launching tubes for the rockets were fitted to a ladderlike frame, which could be elevated to provide the desired range. Once the angle was established and the launcher fixed in elevation, the launching boat was to be rowed to the desired range from the target. Then, the tender would come along side and already prepared rockets would be transferred. The rocket would be placed in the tubular trough at the top of the frame with the cap over its nozzle removed. A flintlock firing device on the trough would then be primed and cocked for release by a long lanyard leading to the rear of the boat. The rocket artillerymen were protected from flames and sparks by heavy tarpaulin cloaks.

Against the French ships nestled in Boulogne harbor the British planned to launch five thousand of the 8-pounder rockets, most of which would carry a 3-pound charge incendiary head. Some, however, would be fitted with the new Shrapnel heads to discourage French crewmen from fighting the fires started by the incendiaries. They were to be fired in salvos of six from twelve rocket boats, each of which would be supplied with forty-eight rounds.

Because of bad weather the rocket attack could not begin until November 21, 1805. On that day Congreve planned to have the launches towed in by brigs. An hour before the attack was to begin, the weather worsened and the wind shifted and increased. Yet the British persisted, moving the rocket launches in as close to the French ships as possible. They were able to fire only two hundred rockets before weather conditions obliged them to pull back. Five of the rocket boats were sunk in the gale.

Lord Keith unhappily observed that "the rockets were without effect; some of them burst in our own boats and none went in the intended direction." The tone of a less biased naval officer was softer: "At first discharge flames broke out in some of the houses and several vessels of the flotilla were fired. Owing, however, to the smoke from the conflagration, it was difficult to see clearly and after the second discharge most of the rockets went over." The fact was that the first military engagement with Congreve war rockets was far from successful.

Inclement weather in the Channel ruled out any further attempts to attack Boulogne in 1805. Revised planning began early in 1806, only to be shelved temporarily as peace negotiations with France were initiated in March. When nothing came of these, preparations continued, with August 20 being set as the date of the second rocket bombardment of the French at Boulogne.

Fleet-leader Commodore F. W. C. R. Owen adopted a drill and set of firing commands for the rockets to be used within his squadron and had them distributed to his ships. One of his disagreements with Congreve was on the mode of launching the rockets. Owen felt that volley fire was inadvisable, while Congreve, who understood the vagaries of his weapon, knew that to be effective rockets would have to be fired in large numbers and by the volley. The larger the volley, the better, especially when the launcher was bobbing about.

The operations order for the attack, sent out by Owen on August 19, stated that "It is intended to attempt the destruction of the enemy's flotilla at Boulogne by means of the fire rockets invented by Mr. Congreve." Upon receipt of the signal 151, the gun brigs were to

> fix the rockets to the sticks, but are not to step the masts for the frames . . . till after dark. When I show two lights abreast . . . two boats will be anchored in the stations from whence the rockets are to be thrown . . . The brigs are to sail around the boats fixed at these stations and each in succession is to fire her rockets in passing the boat . . . They are to use three frames and four rockets in each.

Signal 152, addressed to the rocket launches, meant that

> two rockets only are to be fired from each boat at one discharge . . . Every boat which is fitted to use a rocket frame shall be attended by another boat carrying ten or twelve spare rockets ready fixed and covered with tarpaulins . . . It is proposed that the rocket boats shall anchor with a spring to point the rockets . . . when the boats are placed and the commanding officer gives permission, then discharges are to be renewed as rapidly as possible until the whole are expended.

In this order, Owen partially compromised with Congreve on the mode of employing the rockets. He also added a rather strange afterthought for the crewmen in the rocket boats: "A brave man does not require liquor to support his courage. . . ."

The British fleet was ready to move out of Dungeness on the evening of August 20, but Congreve failed to turn up. A messenger was sent to find him, but by the time he located Congreve, the fleet had received word that peace negotiations with the French were once again under way. The attack was immediately canceled.

September came and went. Peace talks continued and Boulogne was left undisturbed. On September 29, Lord Grenville replaced Howick as First Lord of the Admiralty. Although Grenville was familiar in broad outline with the plans for the "annoyance" of Boulogne, Congreve lost no time in giving him more details. Thirty vessels were to be fitted to launch rockets. The 38- and 36-gun frigates would each carry seventy-two of the 24-pounder rockets and ninety-two guiding sticks. One third of the rockets would have Shrapnel heads and all would be fired from four launcher frames. The 32- and 28-gun frigates would each carry sixty of the 24-pounder rockets and seventy-two sticks. Each would also have three launcher frames. The sloops would carry forty-eight of the 24-pounder rockets and sixty sticks but have only two launcher frames. The gun brigs would be armed with forty-eight of the 42-pounder rockets and an unspecified number of sticks, which were to be fired from three launcher frames. Obviously, there expected to be a certain amount of stick loss in the logistics and preparations for fire.

Peace negotiations broke down altogether in late September 1806, so the British fleet prepared to move on Boulogne. Arriving some four miles off the city on the afternoon of October 8, the rocketeers made a few trial firings and were satisfied that their launching frames had withstood the voyage from England. Rockets were then loaded into the launches and their tenders. By eleven o'clock that night all was ready.

The target of the attack, principally the French transport ships packed tightly in the flotilla basin, escaped damage altogether simply because the rocket boats had come too close in to shore. Since the launchers were fixed in range, there was no means to adjust them once the firing started. Many others went too far to the eastward because the rocketeers had trouble distinguishing their targets in the dark. The firing kept up until five or six o'clock in the morning and all the rockets were expended. An officer aboard the *Clyde* commented that "The bombardment began before three in the morning and continued for more than three hours. Seventy or eighty rockets were distinctly seen at one time in the air by the ships outside." Shortly after the engagement Owen reported to Keith:

> A great confusion among the enemy seemed to follow the first discharge and a fire very shortly broke out which raged in great violence for three or four hours, as we judge, upon the quay near the eastern jetty of the harbour . . . Two things I consider proved by this experiment: the efficacy of the weapon; and the impracticality of using gun brigs here with them.

On October 12 Congreve turned in to Grenville a full account of the event:

> The force decided upon by the Commodore [he wrote] consisted of twenty-four six-oared cutters belonging to the squadron, each carrying a frame for discharging two rockets at a time . . . After a few rounds it was discovered that the place [the French base] was on fire, and the persons who are best informed as to the situation of the different buildings are of the opinion that the fire began in the barracks but that several parts were in flames at the same time & that it burnt fiercely in that quarter where the principal storehouses were known to be . . . They are likewise of the opinion that some of the shipping must have been burnt, as the fire was in the direction of the harbour tho' not of the Basin . . .

Congreve also enclosed a sketch map of the action and showed why the attack failed to materialize as planned. Apparently, the gun brigs could not operate, with the result that the 24-pounder rocket was the heaviest arm fired at Boulogne. "That the vessels in the Basin have not been set on fire," he explained, "is not, I am convinced, from the rockets not having range enough to reach them; but from a sufficient number not having been fired exactly in the direction for them in the late affair."

From this same report to Grenville, we learn that "about four hundred" of the 24-pounder rockets were fired in less than half an hour. Owen, on October 13, wrote to Keith that he learned from some Frenchmen "that some rockets and shells fell on board their vessels on the nights of the 8th and 9th inst. and that two brigs were sunk in the harbour by the latter." It remains doubtful whether or not the rockets sank any vessels in the 1806 attack on Boulogne.

A projected attack on Calais went awry when the mortar brig *Fearless* prematurely opened fire and divulged the British presence. The French returned the fire hotly, and there was no opportunity to put the rocket vessels into play.

Grenville was unimpressed by events of October 8–9 at Boulogne. He commented in a letter to Keith on October 25: "I cannot help thinking that our undisputed naval superiority should give us better means of perplexing our enemies' coasts than by the boat attacks of rockets."

As disappointing as the attacks on Boulogne were, Congreve remained undeterred. Convinced that experience would soon show the worth of his invention, he pleaded that his rockets be used in planned campaigns against the Turks in the Dardanelles (some were actually sent there but arrived too late for action) and against the Danish fleet at Copenhagen. British intelligence had come across details of a secret article in the Treaty of Tilsit, signed between Napoleon and Czar Alexander on July 7, 1807, calling for a joint French-Russian attack on England and for the integration of Danish, Swedish, and Portuguese fleets into Napoleon's forces. In order to block this planned entente, the British decided to seize the Danish fleet, since it was the largest and closest target.

Sixteen ships of the line and frigates sailed for Copenhagen from the Nore, with Congreve accompanying the Royal Marine artillerymen who were to man the rocket boats. This contingent, together with other ships, entered the Øresund on August 3 and ceremoniously exchanged gun salutes with the Danish castle at Kronborg. By August 12 sixty-five British warships were anchored in Elsinore Roads. Almost 28,000 English troops were also on hand.

When the Danes refused to hand over their fleet as demanded, British troops, on August 15, made a landing at Veldhoek, between Elsinore and Copenhagen. Covered by twenty launches converted to rocket boats, they beat off Danish gunboats sent out to spoil the landing.

The British siege works before Copenhagen required twelve days to erect because of a problem in logistics that has plagued amphibious operations throughout history. It is best summed up by a British artillery officer who observed, while drumming his fingers and waiting for ammunition for his land battery: "We are distressed by so many different things being put in the storeships: the things at the bottom are required first; in many instances we have to unload the ship to get at them."

As the final investment of the city was being made, the Danes made two sorties, one on August 17 and the other on August 23. The latter was accompanied by a sharp naval engagement in which the British rocket boats gave covering fire for the counter-attacking gun brigs and sloops. They also engaged the Danish land battery at Trekroner. For four hours the fighting continued, with several of the larger British vessels being forced to withdraw because of damage. The performance of the rockets on this occasion was disappointing principally because of difficulties in finding the correct range. A final naval attack by the Danes on August 31 was easily repulsed.

The bombardment of the city of Copenhagen opened at 7:30 P.M. on the evening of September 2. Among the first rounds to fall in the besieged city were rockets from the launches that had been towed in close to shore. Within five minutes after these initial rockets landed, fires sprang up in three places within the town. The shelling continued without pause until eight o'clock the following morning, when the British land batteries had to be resupplied—with ammunition stored in the bottom of the tender ships. This pattern continued until September 5, when the Danish governor sent out a flag of truce for a parley.

By this time the city was three-quarters burned out, largely by the rockets. Over three hundred houses were leveled. Accounts of the number of rockets fired vary considerably. A letter in a contemporary British newspaper says that "in the three days' bombardment 40,000 rockets were expended." Other observers say between 20,000 and 25,000. While the actual number may never be known, the siege of Copenhagen proved conclusively to the British the value of the rocket as a military weapon.

The Copenhagen campaign was hardly over when Congreve set to work improving the ship-to-ship applications of his fiery weapons and convincing the Admiralty of their efficiency and destructiveness. His diligence was rewarded when, in March

The bombardment of Copenhagen on September 2–5, 1807, with Congreve rockets. (From Schultz, *Danmarkshistorie*, Copenhagen, 1942, Vol. IV, p. 260)

1809, Lord Gambier received word from his superiors that "Mr. Congreve is under orders to join your Lordship with a coppered transport containing a large assortment of rockets and supplied with a detachment of Marine artillery, instructed in the use of them and placed under Mr. Congreve's orders."

The action took place in the Aix Roads near the port of La Rochelle during the evening of April 11. The *Cleveland* came in with 1,300 rockets, supplying them to fireships (in whose rigging they were mounted) and to the cutters *King George* and *Nimrod*. The fireship attack failed—the vessels missed their targets; but the confusion and panic that reigned during the attack caused the French to run their ships aground while attempting to escape from the devilish rockets. The battle lasted sporadically until the twenty-ninth, at which time Gambier informed the Admiralty that he had "every reason to be satisfied with the artillerymen and others who had the management of [the rockets] under Mr. Congreve's direction."

During the next few years the British used Congreve rockets at actions in the Scheldt estuary and Walcheren island in an attempt to destroy Napoleon's dockyards and shipping there; in Spain; at the battle of Leipzig; during the siege of Danzig; and at Waterloo. One French general protested the use of the English rockets as a "monstrous outrage," while another officer told how "Congreve's rockets blazed about in horrible splendour. They are certainly more effectual than shells of any dimensions." In the Leipzig campaign of 1813, one errant rocket went astray and landed among friendly forces causing much confusion. Later, after the proper range had been established, an officer enthusiastically reported that "the effect of the rockets was truly astonishing and produced on the enemy an impression of something supernatural." (The British would take advantage of this effect for years to come.) Captain Richard Bogue later described to the Duke of Wellington how his brigade had employed rockets

> on proper occasions in salvoes projecting 20 shells and case shot or even more; and in favourable ground even double that number, a mass of fire which . . . must be productive . . . of the greatest physical and moral effect, both from the novelty of the weapon, the extraordinary and appalling noise accom-

panying its flight from the first moment of ignition to that of the explosion of the projectile, and from its visibility in flight.

During the Napoleonic Wars, as in earlier wars in India, rockets would often be launched not through the air but along the ground. Tactics calling for the defense of a pass or the covering of a retreat would involve as many as a hundred up to five hundred rockets laid in batteries on the ground and fired in volleys. A Congreve 32-pounder passing along relatively smooth ground could attain a range of between 1,200 and 1,500 feet. Such tactics were very successful against cavalry.

While Bogue and his rocket brigade were participating in the battle of Leipzig, a rocket detachment of thirty-three men under the command of Lieutenant Robert Gilbert arrived at Danzig on August 6, 1813; their purpose was to reinforce Russian and Prussian troops serving under the Duke of Württemberg, who was besieging remnants of Napoleon's army that had withdrawn into the city after the defeats in Russia. During the battle, which continued into November, one defending Frenchman lamented that "the assailants fired a large number of rockets from most of their advanced posts. Shells fell in the city and also between 1,500 and 1,600 rockets. Twenty-two storehouses and barracks, as well as a number of private houses, were set on fire and destroyed . . ." Despite intense bombardment, the French held out until November 29 before surrendering.

Lieutenant Gilbert and his rocketeers were left in their quarters at Kalupki until September of the following year, completely forgotten by the British. When they returned to England on September 10, 1814, they were disbanded. But the Russians were grateful for their services and made Gilbert a Knight of the Imperial Russian Order of St. Vladimir.

Friends and foes alike often expressed their distaste for Congreve's rockets. The Duke of Wellington received them coldly and spoke in terms of horror of the use of these rockets as weapons. On the other side, during a campaign near Bayonne in southern France, a French noncommissioned officer—whose coattails had literally been burned off by a rocket—exclaimed "My God, I've had twenty years of service and I have never seen firearms like those!"

In April 1814, British forces received word that an armistice had been reached with the French (Napoleon abdicated on April 14, and on May 4 he arrived on the island of Elba). A month later, William Congreve succeeded his father as comptroller of the Royal Laboratory and superintendent of the Royal Military Depository at Woolwich Arsenal. In early June 1814 the British began to restage their forces from Europe to North America where a desultory war had been going on since 1812. As a British fleet under Rear Admiral Pulteney Malcolm sailed westward to make rocket-supported attack on the east coast of the United States, a London newspaper trumpeted that "the truculent inhabitants of Baltimore must be tamed with the weapons which shook the wooded turrets of Copenhagen."

Actually, some rockets had been fired earlier. On April 6, 1813, a British squadron under the command of Commodore Sir John P. Beresford appeared off the coast of Delaware and sent an ultimatum to the village of Lewes on Delaware Bay: the Americans were to supply his squadron with food and other provisions—and receive the current Philadelphia prices for them—or face bombardment. The American commander, Colonel Samuel B. Davis, accepted the challenge. When the British attack was over, it was discovered that the rockets had flown over the town and that the bombs had fallen short. Casualties were light: a chicken was killed outright and a pig suffered a broken leg!

The British fared better in Chesapeake Bay at Point Concord and Havre de Grace, Maryland, the next month. Jared Sparks, a witness of the action, wrote:

> Congreve rockets began to be thrown from the barges, the threatening appearance of which produced a still greater agitation, and when one of the militia was killed by a rocket, it was a signal for a general retreat. They left their ground, and escaped with great precipitation and disorder to the nearest woods, even before the enemy had landed.

A couple of days later, the British forces moved up the Sassafras River and sacked Frederickstown and Georgetown after a "terrible discharges of rockets and great guns."

The British employed rockets from one end of the war front

to the other, from Louisiana north to the Canadian frontier. They first used rockets in Canada on March 30, 1814, at a site along the Little Colle River, an offshoot of the Richelieu. Two hundred British, backed by rockets, forced 4,000 Americans to retreat to Plattsburg.

Further south, again in Chesapeake Bay, British naval forces pursued a collection of American gunboats and barges commanded by Commodore Joshua Barney. In an engagement on June 10, 1814, the British rockets proved unnerving to the Americans, and Barney reported that "One of the enemy's rockets fell on board one of our barges, and, after passing through one of the men, set the barge on fire." Later in the month, American militia under Colonel Decius Wadsworth holding positions overlooking the Patuxent River began to retreat in disorder when British rocket and cannon fire was directed on them.

The British also successfully fired rockets during the course of the main attack on Washington overland from Benedict, Maryland (as they moved up the western bank of the Patuxent River and westward through Bladensburg), as well as during a diversionary feint up the Potomac River. American General William Winder testified to the U. S. Congress after the battle that British rockets had "passed very close to the heads of Schultz' and Regan's regiments . . . A universal flight of these two regiments was the consequence." A militiaman later offered the opinion "that not one third of their [British] army came into action at all, any further than by amuzing themselves by throwing Congreve rockets at us. They were so strong we had to give way."

Rocket-tender commander Lieutenant John Lawrence of the Royal Marine Artillery, meanwhile, glowed with pride. Rear Admiral Sir George Cockburn was so pleased with the young officer's efforts that he mentioned him in an August 27 dispatch to Vice Admiral Sir Alexander Cochrane: "I remarked with much pleasure the precision with which the rockets were thrown under the direction of the Marine Artillery." And the Secretary of State for War in England heard from General Robert Ross of "the well-directed discharge of the rockets throwing the Americans into confusion." Lawrence's losses: one rocketeer killed and a sergeant wounded.

Following the sacking of Washington on August 24–25, 1814, the British—on September 10—prepared to move on Baltimore from their base on Tangier Island. Although the troops progressed well, they later decided not to attempt to capture Baltimore but rather elected to bombard Fort McHenry with fire from frigates, bomb vessels, and the rocket ship *Erebus*. One man who was impressed by the sight of the rockets was a lawyer from the District of Columbia named Francis Scott Key. He had gone under a flag of truce to Admiral Cochrane to secure the release of a friend, Dr. Beanes, who had been captured at Bladensburg. The British refused to permit him to return to shore until after the attack on Fort McHenry, so Key witnessed the drama unfold through his telescope. He saw the American flag outlined by "the rockets' red glare" and by "the bombs bursting in air."

The flag, 32 by 29 feet in size, had been sewed in Clagertt's brewery; the sight of it inspired Key to write a poem published a week later entitled "The Defense of Fort McHenry." Later set to the tune of an old English drinking song, "To Anacreon in Heaven," it became a popular patriotic air. (It was not until 1931 that, under the title "The Star Spangled Banner," it became the official United States' national anthem.)

The rockets used against Fort McHenry were standard 32-pounder types developed for sea service. They were 4 inches in diameter, 31.5 inches long, and had 15-foot-long guiding sticks. Their warheads weighed 8 pounds and could attain a range of 3,000 yards. These rockets had little effect on the fort, as most fell short of it.

Over a month before the events at Fort McHenry, on August 9 the British bombarded the town of Stonington, Connecticut, because they mistakenly thought its fishermen were supplying mines for use against British ships. Sir Thomas Hardy, commander of the four-vessel force, sent an ultimatum ashore that read: "Not wishing to destroy the unoffending inhabitants residing in the town of Stonington, one hour is given them from the receipt of this to remove them out of town." The town magistrates answered: "We shall defend the place to the last extremity; should it be destroyed, we shall perish in its ruins."

On board Hardy's flagship, *Ramillies*, were "17 boxes of rockets." During the course of the attack, which started at seven o'clock in the evening and lasted for three days, the British "sent on shore 60 tons of metal," including rockets. Later, the *Connecticut Courant* reported: "The Congreve Rockets which were frightful at first soon lost their terrors and effected little." Soon afterward, Philip Freneau composed a short poem commemorating the event:

They killed a goose, they killed a hen;
Three hogs they wounded in a pen;
They dashed away,—and pray what then?
That was not taking Stonington.

The shell were thrown, the rockets flew;
But not a shell of all they threw—
Though every house was in full view—
Could burn a house in Stonington.

As the rocket as a war weapon became better known and its limitations realized, it ceased to terrorize the Americans. Poems were not written concerning its use in the New Orleans campaign in December 1814 and January 1815, but a British lieutenant remarked that, in one engagement, "A few rockets were discharged which made a beautiful appearance in the air, but the rocket is an uncertain weapon and these deviated too far from their objects." An engineering officer from Louisiana, A. Lacarrière Latour, was more accurate when he explained that

> The British had great expectation from the effect of this weapon, against an enemy who had never seen it before. They hoped that its very noise would strike terror into us; but we soon grew accustomed to it, and thought it little formidable; for in the campaign, the rockets only wounded ten men, and blew up two caissons. That weapon must doubtless be effectual to throw amongst squadrons of cavalry, and frighten the horses, or to set fire to houses; but from the impossibility of directing it with any certainty, it will never be a very precarious weapon to use against troops drawn up in a line of battle, or behind ramparts.

General Andrew Jackson agreed, as he shouted to his troops,

The Battle of Waterloo on June 18, 1815, showing bombardment by Congreve rockets. Contemporary aquatint.

"Pay no attention to the rockets, boys; they are mere toys to amuse children!" After the battle of New Orleans, he praised the valor of his heterogeneous army of regulars and militia that had defeated the English: "The enemy has retreated . . . their rockets illuminated the air . . . the glare of their firework rockets . . . served only to show the emptiness of his parade and to inspire you with a just confidence in yourselves."

In Europe, English rocket troops once again engaged Napoleon's forces, from the time the temporarily deposed emperor landed near Cannes on March 1, 1815, through to his final defeat at the battle of Waterloo on June 18 of that year. In one skirmish, a French soldier cursed the weapons, complaining that "the English were trying to burn them [the French troops] alive." On another occasion, the Second Rocket Troop under Captain Whinyates ran up against a French Old Guard square and helped thrash it smartly. After the battle of Waterloo, his troops marched with the Allied army into Paris where they remained with the occupation forces until ordered back to Woolwich Arsenal toward the end of January 1816.

With the threat of Napoleon gone and the British withdrawn from the United States, the Prince Regent of England reduced the forces both of the Royal Horse Artillery and of the rocket service. In the transition into peace, the manufacturing facilities at Woolwich Arsenal were closed on October 26, 1818, the rockets on hand were placed in storage, and the skilled workmen were laid off with a fortnight's pay. The future of war rocketry seemed hazy indeed.

From 1819 to the end of the nineteenth century, however, Congreve and later Hale rockets were manufactured and fired in limited quantities. The most popular versions were the 12-pounder with a Shrapnel charge of forty-eight lead balls and a 2,500-yard range; the 32-pounder carrying incendiary, Shrapnel or fragmentation warheads from 2,000 to 3,000 yards; and the 3,500-yard 42-pounder available with a similar choice of warheads.

Sir Thomas Cochrane the former scourge of Chesapeake Bay, secured for himself some rockets after the close of the War of 1812. Now unemployed—"in consequence of my unjust expulsion from the British naval service, by the machinations of a powerful political party which I had offended"—he offered his services to the Chilean independence movement of Bernardo O'Higgins. A rocket-manufacturing laboratory was set up at the Santiago arsenal under a former Congreve employee named Goldsack, but unfortunately, half the workers were Spanish prisoners of war, who wasted no time in sabotaging the weapons.

Cochrane then mounted his rockets on towed rafts outside the harbor at El Callao, Peru, and proceeded to bombard the Spanish fleet anchored there. The effect was, he said, "unfavourable." Cochrane explained that

> Great expectations were formed . . . as to the effect to be produced by these destructive missiles, but [we] were doomed to disappointment [as they] turned out utterly useless. Some, in consequence of the badness of the solder used, bursting from the expansive force of the charge before they left the

Setting up rocket launchers previous to attacking stockades in Rangoon, Burma, on July 8, 1824. Contemporary print.

> raft, and setting fire to others—Capt. Hind's raft [from the ship *Araucano*] being blown up from this cause, thus rendering it useless, besides severely burning him and thirteen men: others took a wrong direction in consequence of the sticks not having been formed of proper wood, whilst the great portion would not ignite at all from a cause which was only discovered when too late.

The "cause," it soon turned out, was the Spanish workers, who had been used "from motives of parsimony," as Cochrane put it. They "embraced every opportunity of inserting handfuls of sand, sawdust, and even manure, at intervals in the tubes, thus impeding the progress of combustion." As for Goldsack, he was soon "overwhelmed with reproach for the failure of his rockets." The first post-Napoleonic War use of rockets was hardly auspicious.

Yet rockets continued to appear. During the Burmese conflict of 1824–25, for example, the Burmese commander Maha Bandoola promised his king that he would capture Calcutta and carry off the British governor general in golden chains. But after taking Rangoon, Major General Sir Archibald Campbell led a combined force of British and East India Company troops up the Irrawady River and upon making contact with the enemy, unleashed rockets. Bandoola was killed and the Burmese troops fled.

In a similar action along the Gold Coast in Africa, Ashanti warriors panicked at Dordowa near Accra as Congreve rockets roared into their ranks. During the Civil War in Spain that ended in 1840, the Miguelites were driven back from Pedroite positions by British mercenaries who had armed themselves with rockets. One report told of troops on "the left of the lines" being "followed by Congreve rockets, which, although not doing . . . much damage hastened their retreat."

A Chinese minister named Keshen reported to his emperor during the course of the Opium War in the early 1840s that "it appears to your majesty's slave that we are very deficient in means, and have not the shells and rockets used by the barbarians. We must, therefore, adapt other methods to stop them, which will be easy as they opened negotiations." But talks got nowhere, so the British moved against Canton, unleashing rocket fire from positions near the village of Tsingpoo. The Chinese yielded.

The story was the same during a punitive raid against Argentina by British and French ships operating in the Rio de la Plata in 1846, where rocketeer commander Lieutenant Lauchlan B. Mackinnon wrote: "It is quite impossible to describe the panic and confusion this [the fire of the rocket battery] caused amongst the enemy, as it was the first intimation of any attack from the island. Suffice it to say, the whole space was cleared in a moment." The enthusiastic young officer remarked that fire from his six rocket tubes "was equal to a continuous discharge from the artillery of two 80-gun ships." He claimed that he could get off forty rockets a minute.

The last major conflict of the nineteenth century in which rockets were used by all combatants was the Crimean War (1853–56). The British employed essentially the same rockets as they had during the Napoleonic wars forty years earlier. A total of 382 were logged as having been fired by them during the siege of Sebastopol and many more during the defense of Eupatoria. Royal Marine rocketeers also participated in a series of naval

Russian war rockets firing from a window during the Crimean War, 1853–56. (From USSR Ministry of War, *Artilleria i Rakety,* Moscow, 1968, p. 44)

raids in the Sea of Azov. In an action against the town of Taganrog, rockets set several grain warehouses afire. They also proved effective in a series of incidents in the Baltic directly associated with the Crimean campaigns, including actions in the Russian ports at Vyborg and Sveaborg in 1855.

A British marine won the Victoria Cross for his bravery at Vyborg on July 13, 1855. The Russians were firing rockets against the cutter of the *Arrogant* when a spark from one ignited an English rocket aboard the cutter. It went off instantly, piercing a hole through the boat. All hands jumped overboard, and three men subsequently drowned. Lieutenant G. D. Dowell of the *Ruby* was nearby supervising the resupply of his ship's rockets when he saw what had happened. He immediately put the *Ruby*'s quarter boat over the side and with three volunteers went after the stricken men and cutter. He rescued the men and then returned to bring in the cutter to prevent it from drifting into shore and being captured. At Sveaborg, on August 9, British rocket launches provided heavy fire support to the general bombardment, setting off many fires and adding to the general confusion.

From 1857 to the end of the century, the British used rockets in sporadic colonial campaigns. Rocket "cars" (small, rocket-powered wheeled devices) were used effectively during the Indian Mutiny of 1857, one English officer remarking: "The very sight of the little car, with a mast stepped in its centre, made your hair stand on end. A more diabolical apparatus for rousing an enemy has never been invented. Their effect must have been terrific."

The newly adopted Hale rocket was issued to British artillerymen for action in the Abyssinian campaigns of 1867 and 1868, often referred to as "King Theodore's war." Among General Robert Napier's 13,000-man forces was a rocket battery of the Naval Brigade armed with four launchers and a supply of 340 Hale 6-pounder rockets. Their objective was to rescue two British civil servants and fifty-eight other European hostages held by King Theodore partly because of his approaching madness and partly as a result of his anger at Queen Victoria who had declined to answer a letter he had sent to her in 1864.

Rocket boats from the British war vessels *Harrier* and *Cuckoo* destroying Russian shipping at Nystad, Finland, during the Crimean War. Contemporary engraving.

Napier's forces, coming from India, disembarked on the coast of Eritrea on the Red Sea and, in the face of extreme logistic difficulties, pushed into the Ethiopian highlands. Hale rockets went into action at the battle of Arogee on April 10, 1868. When one of the hissing missiles barely missed him, Theodore flung his shield up over his head and cried, "What a terrible weapon! Who can fight against it?"

Theodore committed suicide after the battle, and following the fall of Magdala three days later, the hostages were released.

In Africa the Ashanti chieftains expressed similar feelings about rockets during the war of 1873–74. In one engagement thirty-six Hale 9-pounders were brought into action under the command of Lieutenant A. Allen against an Ashanti village. One British officer later wrote of the "screeching of the rockets . . . all this for half an hour made up a scene it is impossible adequately to depict." Later, when some 10,000 Ashantis attacked British marines at Dunkwa, rocket power saved the day. Because of the Ashanti fear of rockets, General Sir Garnet Wolseley ordered that all advances by the British were to be "preceded by a heavy fire of guns and rockets."

Toward the end of the century the British were at war with the King of Benin (Benin later became the country of Nigeria).

Rear Admiral H. H. Rawson developed plans for amphibious attack on the capital of Benin. The march was difficult. Natives fired on the advancing column from the bush. Waterholes were found dry. Malaria struck. The heat was oppressive. Along the route, the bodies of sacrificed natives were found; they were to form a "juju" (a magic barrier) which the British were not supposed to be able to cross. Nevertheless, on February 18, 1897, the British reached to within a mile of the capital and set up their rocket launchers. By chance, a few rockets landed within juju compounds of the natives, causing instant panic. The capital fell without resistance.

In the Sudan meanwhile, after the massacre of General Charles Gordon and his garrison at Khartoum on January 26, 1885, the Sudan fell under the control of the dervish forces led by the Mahdi Mohammed Ahmed. When the campaign to reconquer the Sudan began in 1896 under General Herbert Kitchener, British rocket troops again saw action.

In one incident, the crew from the gunboat *Fateh* under Lieutenant David Beatty, R.N., unleased their rockets against the fortified dervish camp at Nakheila on the south bank of the Atbara River, about thirty miles from its confluence with the Nile. When their huts began to burn, the dervishes abandoned their camp, which helped open the road for Kitchener's Anglo-Egyptian army pushing toward Omdurman and Khartoum. After another action, a British marine sergeant related that

> Several days were employed in trying to coax the enemy out into the open, but without success; so the Sidar determined to attack, and finding by reconnaissance that they had constructed numbers of huts of dry grass in their camp, sent word to Capt. Keppel, R.N., for a rocket party from the gun-boats . . . Camels carried the gear, and we marched. . . . On account of the distance and nature of the ground, I was unable to get in any good work with the rockets . . . [so] we took up a good position at 500 yards range, where I was able to get in some good shots, setting their camp on fire in several places.

With the reconquest of the Sudan from the forces of the Mahdi, the age of Congreve—and of Hale and others, too—came to an

The Hale rocket was used by the British expeditionary force during the Abyssinian campaign of 1867–68 (also known as King Theodore's War). After the war was over the rocket was demonstrated at Senafe, as shown here, for Kassai, a local prince and ally of the British. The rockets were fired by gunners of the Naval Brigade. Contemporary print.

end for England. Also during the nineteenth century, a chronicle of Russian, French, Italian, Danish, Swedish, and other national rocket programs would parallel and reflect British experience. Powder rockets had gone as far as they could without the application of a sophisticated guidance and propulsion technology that would not become available for decades.

Now, let us skip ahead once more across the intervening decades, as the scene shifts from the Congreve and Hale brand of rocketry to that of a new breed of experimenters. The next chapter in rocket history was to be written not in England, France, Russia or other European countries that had shown so much prowess in the nineteenth century, but in post-World War I Germany.

Winter, 1944 V-2 launch test.

7

PIONEERING MODERN ROCKETRY

Victor Hugo said that there is no stronger force in this world than an idea whose time has come.

During the waning years of the nineteenth century and the opening decades of the twentieth, a new breed of dreamers and inventors began examining the feasibility of one of the most grandiose and exciting ideas in human history: manned travel beyond the earth's atmosphere. The basis of their studies was the lowly rocket, the only propulsion system that their theoretical investigations revealed could operate in the near-vacuum of outer space. The time had arrived for an idea to merge into reality, as four outstanding men independently evolved sound concepts of how space travel could come about.

These four men, who lived in Russia, the United States, Germany, and France, applied their intellectual wizardry and experimental prowess to bring the requirements for a real breakthrough in human progress into clear focus and to catalyze the latent potential of the rocket into a concerted endeavor that was to culminate with the landing of Apollo 11 astronauts on the Moon. Though they worked apart from one another, Konstantin Eduardovich Tsiolkovsky, Robert Hutchings Goddard, Hermann Oberth, and Robert Esnault-Pelterie had something in common. Their imaginations were inspired by the great nineteenth-century writers of imaginative fiction, Jules Verne and H. G. Wells—men who made space travel sound exciting and, even more important, whose ideas sounded technically feasible to young boys with an aptitude for science and engineering. These two elements of what

As the rocket became better known during the course of the nineteenth century, many ideas—some sound, most fanciful—were proposed in the effort to find a way for rockets to propel man into the

air and even into space. They ranged from the curious Golightly steam rocket to Battey's aerial ship, shown here. (Courtesy Maria Cooper Janis)

we now call science fiction literature spurred serious minds toward serious scientific problems.

How many persons for how many years had idly watched sky-rockets streak into the air and thought nothing of it except to marvel at the colorful pyrotechnics? With the magic of Verne's prose in his mind, Tsiolkovsky began as early as the 1880s (he was born in 1857 in Izhevskoye) to ponder what it one day might accomplish:

> For a long time I thought of the rocket as everybody else did [he later recalled]—just as a means of diversion and of petty everyday uses. I do not remember exactly what prompted me to make calculations of its motions. Probably the first seeds of the idea were sown by that great fantastic author Jules Verne—he directed my thoughts along certain channels, then came a desire, and after that, the work of the mind.

Nearly two centuries elapsed between Newton's formulation of the laws of motion and Tsiolkovsky's mathematical proof that the rocket was the only means by which man would some day place himself into space. (The Russian pioneer predicted this occurring in the year 2017; he was in error by fifty-three years.) His mathematics revealed the principle of mass ratio. This told him that the performance capability even of a rocket of immense size and weight was limited. His formula also suggested two methods how rocket performance could be maximized. He had to find the best combinations of propellants to increase the velocity of his exhaust gases, and he had to reduce the empty weight of the rocket and all its parts in order to carry more propellants.

These two avenues opened up new vistas for theoretical research, the most important of which was to determine what combinations of fuels and oxidizers would produce the highest exhaust velocities. Tsiolkovsky's calculation, done painstakingly by hand, indicated that the best propellants for practical use were kerosene and liquid oxygen or liquid hydrogen and liquid oxygen. He even pointed out that ozone as an oxidizer would be better than liquid oxygen. In this suggestion, he is still ahead of contemporary propellant technology.

Other studies led him to the principle of rocket staging as a means of achieving the velocities needed to escape from Earth's gravity. He realized staging could be done in two ways: in series or in parallel. One rocket could be placed on top of another as the early artificers had done and as illustrated in more modern times by the giant Saturn 5 Moon rocket, or several could be clustered into a bundle as we do today with the Titan 3C.

Tsiolkovsky's deep insight into the full scope of astronautics led him into many nonrocket areas, such as working out the sort of life-support system future astronauts might require. The implications of his pioneering work are obvious to the space community today. It supplied man with the mathematical tools needed to design multistage launch vehicles. Moreover, his studies of propellant chemistry and rocket propulsion established the starting point for the design of modern rocket engines such as Saturn 5's F-1 and J-2 models. These engines are powered by liquid oxygen and kerosene and liquid oxygen and liquid hydrogen, respectively, proving that Tsiolkovsky's findings of many decades ago are as valid today as they were then. His theories have stood the tests of time.

Contemporary military and space rocketry also owe their debt to the American Robert H. Goddard, a shy, brilliant professor of physics at Clark University who worked during the opening decades of our century. Nearly thirty years younger than Tsiolkovsky, Goddard was certainly his intellectual peer. The American was a theoretician and a teacher like the Russian, but he was also a builder. He was the perfect example of the practical New Englander who likes to prove things to himself. His work most relevant to modern rocketry began shortly after World War I, but his theoretical studies and experiments with gunpowder rockets antedate that great conflict.

Goddard's monograph *A Method of Reaching Extreme Altitudes,* published by the Smithsonian Institution in 1919, is a classic in the literature of astronautical science. It was a primer for the early research in rocketry in the United States and became well known in Europe. Yet Goddard was not content with mathematical models and theories. He wanted to design, build, and fly his own rockets.

So he did. During the many years of his active life, and even posthumously, he received more than two hundred patents in the field of rocketry alone. He proved by experiment in the field

Dr. Robert H. Goddard holds barograph in its case. This will be placed in the cap suspended by rubber links under tension. Inside the cap is shown the bag for the main rocket parachute. In an assistant's right hand is the bag containing the small parachute for the cap. The small pilot chute for opening cap chute is shown (white package) below. Shown also is the loop of strong steel cable by which the large parachute is attached to the main part of the rocket. Roswell, New Mexico, 1940. (Courtesy Esther Goddard. Photograph by B. Anthony Stewart, copyright National Geographic Society)

that liquid-propellant rockets could be built and flown just as he and Tsiolkovsky before him had mathematically predicted. He introduced the gyroscope control system to rocketry, turbopump-fed liquid-propellant engines, regeneratively cooled (that is, cooled by one of the rocket's own propellants) engines, and gimbal-mounted engines (permitting the rocket to be steered like an outboard motorboat). His first four liquid-propellant rocket flights were made near Worcester, Massachusetts, on

March 16 and April 3, 1926, December 26, 1928, and July 17, 1929. The following year he moved his flight testing activities to a site near Roswell, New Mexico, where he continued to work for many years.

As early as 1912, the French pioneer Robert Esnault-Pelterie began lecturing on the principles of space flight to such prestigious organizations as the Société Française de Physique. His *L'Exploration par fusées de la très haute atmosphère et la possibilité des voyages interplanétaires,* published by the Société Astronomique de France was a solid ninety-eight-page book that examined the possibilities of exploring interplanetary space by rocket. Esnault-Pelterie later wrote *L'Astronautique* (1930) and *L'Astronautique–Complément* (1934), works that thoroughly reviewed the potential of astronautics.

The fourth of the original pioneers of modern astronautics was Hermann Oberth, the youngest and the only one to witness the fruition of his early theoretical endeavors. (At the age of seventy-five he personally watched the launching of Apollo 11 at the Kennedy Space Center, Cape Canaveral, Florida, on July 16, 1969; five years later he read how Skylab space-station crews logged more than five hundred man-days in space before that epoch-making mission terminated on February 8, 1974.) Fate dictated that the theoretical work Oberth started and the individuals inspired by his visions would have far-reaching influence on the future. For this reason and this alone, we must follow in some detail the episode of his life and of the events that it engendered.

As so often occurs in history, great achievements become possible only if the political, military, and intellectual climates are propitious. Oberth, though he certainly did not realize it when he started out on a trail that led to the Moon, lived in the right geographical area at the right time in history. More a theoretician like Tsiolkovsky and Esnault-Pelterie than a designer and builder like Goddard, Oberth was born in 1894 in Hermannstadt, a small town in Transylvania, or Siebenbürgen, a German-speaking region of the former Austro-Hungarian empire which, after the downfall of the Hapsburgs, became a part of Romania. In 1923 his trail-blazing *Die Rakete zu den Planetenräumen,* or *The Rocket into Planetary Space,* appeared. Its preface contained four Oberth pronouncements that rang in the age of space travel:

1. With the present state of science and technology, it is possible to build machines that can rise to altitudes beyond the Earth's atmosphere.
2. With further improvement such machines can attain velocities that will enable them to stay unpowered in outer space without falling back to Earth. They may even be capable of leaving the Earth's gravitational field.
3. Such machines can be built so that people can ride them, in all likelihood without any harm to their health.
4. Under certain conditions the construction of such machines may become commercially profitable. Such conditions may come about within a few decades.

Oberth's book with these ringing declarations lit the fire of a running controversy in scientific circles that did not end until, forty-six years later, Neil Armstrong set his foot on the Moon. In its soberly written text, the book offered the first comprehensive scientific treatment of the rocket principle in its application to space flight. In addition, it developed the concept of the liquid-propellant rocket with separate tanks for fuel and oxidizer, and it put hard figures behind all its feasibility claims. Enclosed in the book was a fold-out line drawing of a two-stage, liquid-propellant, high-altitude rocket. The first stage was powered with alcohol and liquid oxygen, the second with liquid hydrogen and liquid oxygen. Guidance was provided by gyroscopes and other inertial references.

In 1929 Oberth came out with a vastly enlarged third edition of his book. It was easily four times as voluminous as the 1923 edition, contained a wealth of additional information, and was published under the new title, *Roads to Space Travel.*

Oberth's many attempts to find a sponsor willing to finance some of the urgently needed experimentation met with partial success when producer Fritz Lang of the German UFA film corporation invited him in 1928 to Berlin to become scientific adviser for his new space-flight film *Girl in the Moon.* Lang, whose fame later brought him to Hollywood, had made quite a name for himself as a producer of spectaculars. He had now become fascinated by the motion picture potential of manned space flight. His wife, Thea von Harbou, had prepared a fine script.

The project gave Oberth a wonderful opportunity to lay out a detailed model for a two-stage, liquid-propellant, manned spaceship in which he could further clarify his thoughts. But it also brought him much grief. For one thing, it greatly troubled Oberth that Fritz Lang insisted on a breathable atmosphere on the Moon. The great producer said he could not film lunar scenes with the actors wearing space helmets, nor could he accept actors floating around in zero gravity inside the spaceship as it would look ridiculous. Even more serious than these minor artistic licenses was the pressure Oberth had from UFA's promotion department. Someone had hit upon the idea that Oberth should build a high-altitude rocket as described in his book and that this rocket should be launched on the day of the premiere of *Girl in the Moon.* Attracted by the long-sought opportunity to build hardware with the funds offered by the promotion department, Oberth unfortunately agreed to try to do his best.

He worked against an impossible schedule. The premiere was only a few months away while Oberth was still trying to set up some elementary combustion experiments with the newly designed gasoline-liquid oxygen motor that was to power the rocket. The experiments were conducted in the UFA studio complex in fashionable Neubabelsberg and, not too surprisingly, the fire marshal soon took a dim view of large supplies of liquid oxygen and long torchlike flames amid the highly combustible stage sets and supply warehouses all around. On October 15, 1929, the premiere took place without the promised rocket launch, but *Girl in the Moon* became a great success nevertheless.

To expedite his experiments at the Neubabelsberg studios, Oberth placed a want ad in the papers for a technical assistant. From the applicants, he selected a man who impressed him with his practical outlook and his positive make-do attitude. Rudolf Nebel, who had been a World War I fighter pilot, had a master's degree in engineering and expressed a strong desire to work in

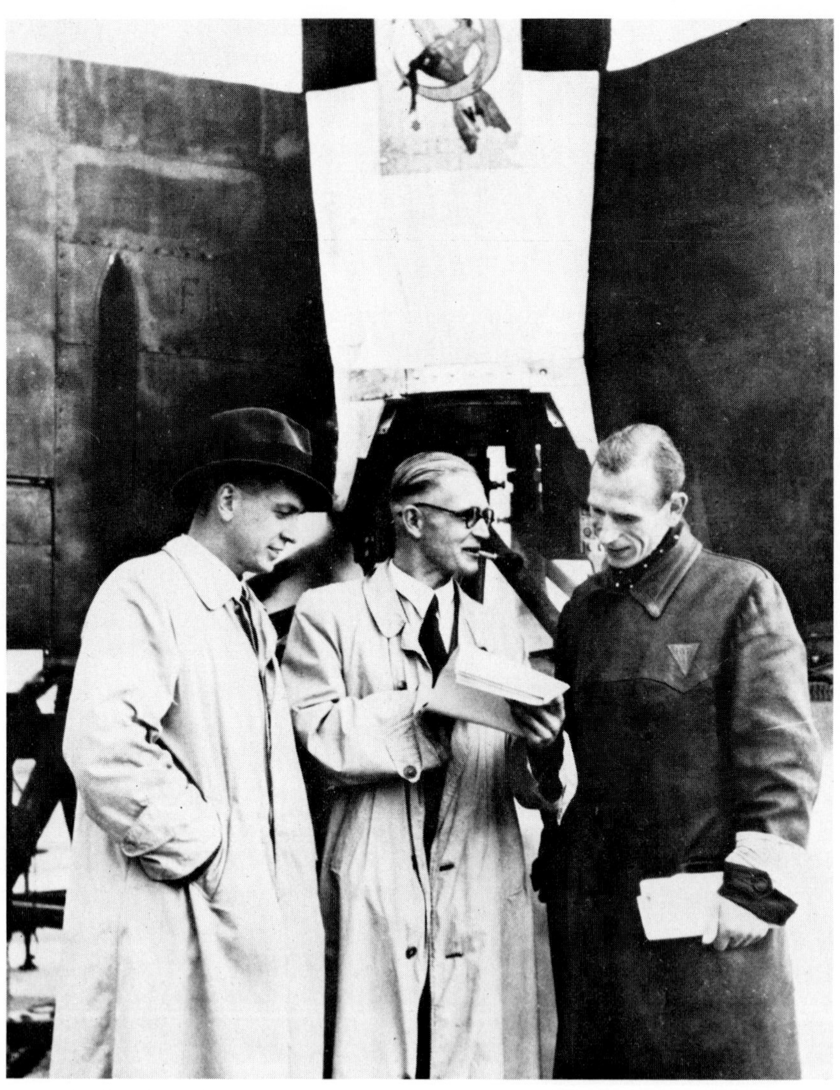

The "Girl in the Moon" film was not forgotten by the German experimenters who, a decade and a half later at the great World War II rocket development center in Peenemünde, painted a girl, a rocket, and a crescent moon on the aft end of their A-4 (later called the V-2) missile. Engineers Werner Gengelbach, Walter Thiel, and Hans Hüter stand in front of the missile.

the experimental field, as the depressed economic conditions had temporarily compelled him to work as an industrial sales representative.

As a first logical step to demonstrate the feasibility of steady combustion of two separate streams of gasoline and liquid oxygen injected into a combustion chamber, Oberth had designed a small rocket motor he called "Kegeldüse." With Nebel's help, the construction of this device and the purchase of supporting equipment made good progress, but because of the missed premiere date the UFA promotion department's interest had cooled to the point that it even refused to pay some of the bills. Oberth satisfied a few debtors by chipping in from his own sparse funds and a 10,000-franc prize he had won for his new book from some French admirers. Finally he returned disgusted and penniless to his Romanian home.

In May 1930 he received word from Nebel that the Chemisch-Technische Reichsanstalt (the equivalent of the chemical section of the U. S. Bureau of Standards) in Berlin-Plötzensee had offered the use of an experimental area and some workshop facilities to set up a series of experiments with the Kegeldüse. Oberth immediately returned to Berlin and was busy the next eight weeks with test preparations. The gasoline was pressure-fed into the Kegeldüse from a cylindrical ten-gallon container. The liquid oxygen was poured into a thin-walled copper vessel inserted into a massive steel tank. Both containers were pressurized from a steel cylinder containing high-pressure nitrogen. The 2,000 psi (pounds per square inch) pressure in the cylinder was reduced to 150 psi by a welder's pressure reducer. The gasoline and liquid oxygen were piped to the Kegeldüse by ¼-inch copper tubing. The Kegeldüse itself, with the exhaust nozzle pointing upward, was submerged into a water-filled bucket for cooling, which in turn was placed on a grocer's scale to measure the thrust. Fuel and oxidizer consumption during operation of the Kegeldüse were measured by gauging the container contents before and after the test.

Two new faces had joined Oberth's fold: Klaus Riedel, a young engineer with the Siemens works, and Wernher von Braun, then a first-semester engineering student at the Berlin

Institute of Technology. Both were members of the German Society for Space Travel which had been formed in 1927 by Johannes Winkler. Upon hearing of Oberth's return to Berlin, they had offered their services free. When they pointed out that while enthusiastically believing in the future of space rocketry, they had absolutely no experience to offer, Oberth calmed their concern with the remark that he himself had no experience either, nor was there probably anyone else who had.

On July 23, 1930, came the great day. The Kegeldüse worked beautifully. For over a minute and a half a fiery jet three feet long roared from the exhaust nozzle. The diamond pattern of interacting shock waves clearly proved that the gas left the nozzle at supersonic speed. Dr. Ritter the head of the Reichsanstalt, certified that the Kegeldüse had performed as follows:

Thrust: 7 kilograms for the first 50.8 seconds, thereafter 6 kilograms for another 45.6 seconds.

Gasoline consumed:	1 kilogram
Oxygen consumed:	6.6 kilograms
Exhaust velocity:	756 meters per second

Unknown to Oberth and his helpers, another group at the opposite end of Berlin had beaten them to the punch. Max Valier, whose solid fuel-powered rocket cars had acquired much newspaper fame, had quietly teamed up with Dr. Paul Heylandt, a successful pioneer in the use of liquefied gases for industrial purposes. Their joint objective had been the development of a liquid fuel-powered rocket engine. Valier's plan was first to use it to demonstrate and popularize rocket propulsion in automobiles. Later he hoped to introduce it to aviation.

This work, supported by several engineers and the fine facilities of Dr. Heylandt's plant in Berlin-Britz, began in a very promising vein. On April 17, 1930, a car equipped with an alcohol-liquid oxygen rocket engine of 20 to 30 kilograms thrust, was tried out successfully at Tempelhof airport. The propellant supply lasted for eight to ten minutes.

Exactly one month later, Max Valier, the indefatigable speaker, writer, experimenter, and promoter of rocket-powered space flight, was dead. During a static test with a new fuel the little motor exploded and a tiny fragment pierced his lung. He died in the arms of his associates, Walter Riedel and Arthur Rudolph, who subsequently carried on his work. Walter Riedel (no kin to Klaus Riedel) later became head of the design office at Peenemünde, Germany's World War II rocket research station, and played a major part in the development of the A-4 rocket that became known as the V-2. Rudolph also went to Peenemünde where he ran some of the experimental shops. After the war he went with Wernher von Braun to the United States and finally became project director for the mighty Saturn 5 rocket that carried Apollo astronauts to the Moon and lofted the Skylab space station into orbit.

Valier, who hailed from Bozen (modern Bolzano) in South Tyrol, and Oberth were not the only rocketry and space-flight prophets born in the old Austro-Hungarian Empire. In 1926 Dr. Franz von Hoefft and Baron Guido von Pirquet formed the Scientific Society for High Altitude research in Vienna. Von Hoefft proposed a generic family of liquid rocket-powered vehicles, ranging from a simple parachute-recoverable meteorological research rocket to modernistic multistage, hydrogen-oxygen-powered spaceships, while Von Pirquet advocated the establishment of an orbital space station and contributed a great deal to the understanding of the fundamental processes in a rocket engine. The space station idea was further refined by Hermann Noordung (a pseudonym for the Czechoslovakian Captain Hermann Potočnik) in his 1928 book, *The Problem of Travel Through Space.* During the same year, Franz Abdon von Ulinski conceived of the "cathode-ray spaceship" which vaguely anticipated our modern plans for solar battery-powered spacecraft with ion propulsion. None of these men, however, designed or built any hardware or conducted experiments.

One other Viennese followed Oberth's and Valier's lead and became a very active and successful pioneer in the development of liquid-propellant rocket engines. Eugen Sänger conducted the first test of his regeneratively cooled rocket motor in January 1933. Three months later he published an excellent book entitled *Rocket Flight Technique,* in which he analyzed in great scientific detail a liquid propellant rocket-powered supersonic aircraft. Stimulated by his advanced concept, the German Luftwaffe began, in April 1937, to construct for Sänger an elaborate rocket propulsion facility at Trauen, near Hamburg, where, until the end of World War II, he pioneered the use of fuel oil/liquid oxygen in high-pressure rocket engines. Among many other things, he also experimented with metal dust dispersed in fuel oil to raise combustion temperature and thus specific impulse.

Finally, in 1931, Friedrich Schmiedl of Graz launched the world's first official mail rocket. His idea of carrying letters with short-range rockets to isolated mountain hamlets persuaded the Austrian postal service to authorize, on an experimental basis, the launching of 102 letters to which special rocket stamps had been affixed in addition to the regular stamps. Although Schmiedl's 50-pound solid-propellant rockets performed flawlessly, the system was not accepted for full-fledged operational use.

Meanwhile, all sorts of activities were going on in Germany. On June 5, 1927, Max Valier had persuaded a bright young student at the Breslau University, Johannes Winkler, to form the Verein für Raumschiffahrt (Society for Space Travel) as a collecting pool for space-minded people. Membership in Winkler's society mushroomed. From 1927 to 1930, the VfR published a first-class monthly journal, *Die Rakete,* with contributions from many leading contemporary advocates of rocketry and space travel.

In 1931 Winkler moved to the Junkers aircraft company in Dessau where he studied the feasibility of solid rockets for assisting heavily loaded airplanes in getting off the ground. On the side, he also designed a meteorological high-altitude liquid-fuel rocket. With financial assistance from some private sponsors, he continued some of the experimental work with a small liquid rocket motor that he had begun in Breslau. He tested his motor in a light-weight rig which he labeled a "flying test stand." On February 21, 1931, this rig rose to an altitude of thirty feet. Recovered undamaged, it was launched again on March 14, 1931, and this time the rig flew through an arcing trajectory with an apex height of 180 feet and crash-landed at a distance of 570 feet from the take-off site. It was the first successful flight of a liquid-fuel rocket in Europe.

This success encouraged Winkler and his assistant, Rolf Engel, to tackle the construction of the meteorological rocket. It was to be a streamlined vehicle with three fins, a total length of 62 inches and a diameter of 13 inches. The rocket was to be loaded with 70 pounds of liquid oxygen and 9 pounds of liquefied methane. The thrust of 210 pounds was to be generated by an uncooled steel motor whose metal mass was to absorb the heat from the oxygen-rich combustion for the 49 seconds duration of the burn.

After four reasonably promising static tests, Winkler took the rocket to a water-surrounded sand spit in East Prussia for a launch attempt. Unfortunately, the Baltic salt spray caused corrosion and leakage of the valves, permitting methane gas to seep into the air-filled spaces between the rocket's tanks and outer skin. At the moment of ignition, the rocket was destroyed in a violent explosion.

Upon Oberth's return to his Romanian teaching post in late August of 1930, soon after his Reichsanstalt triumph, his first assistant, Rudolf Nebel, took charge of the team he had left behind. Not far from the Chemisch-Technische Reichsanstalt, where Oberth's Kegeldüse had been tested and certified, Nebel discovered an abandoned ammunition storage area and succeeded in talking the city fathers of suburban Berlin-Reinickendorf into a free lease for an indefinite period. The site comprised about 300 acres with various abandoned buildings, bunkers, and concrete blockhouses surrounded by blast walls. The whole area was badly overgrown by weeds and underbrush, but otherwise perfect for the purpose. On September 27, 1930, operations were set up in one of the blockhouses, after the mounting of a somewhat grandiloquent sign reading "Raketenflugplatz Berlin."

Nebel's team consisted of the two men who had worked with him at Oberth's Reichsanstalt tests, Klaus Riedel and Wernher von Braun. A few new faces at the Raketenflugplatz included Hans Hüter, Kurt Heinisch, Helmuth Zoike and two journeymen, one Paul Ehmeyer and the other named Bermüller. A separate group of buildings was put at the disposal of Winkler and his assistant, Rolf Engel. Willy Ley, vice president of the Verein für Raumschiffahrt and a most effective science writer, tried to establish a respectable press image of the small enterprise. Finances were practically nil but enthusiasm was immense. The main source of income of the Raketenflugplatz was Nebel's uncanny ability to wangle anything from raw materials to free services and food from those who were susceptible to the lure of space travel. The growing labor force cost nothing by reason of the then prevailing general unemployment. Draftsmen, electricians, sheet-metal workers, and mechanics were only too happy to take up residence rent free in one of the Raketenflugplatz's unused buildings and to maintain their various skills. Soon the staff had grown to about fifteen people.

The first task was to devise a good rocket motor. Oberth's Kegeldüse had been cooled by the stagnant water in the bucket into which it had been submerged. To improve propellant economy, it was necessary to increase both combustion temperature and pressure and this required coolant circulation. In January 1931, Nebel and his key design and test engineer, Klaus Riedel, came up with a welded aluminum motor that was cooled by circulating water and produced a thrust of 40 pounds. On May 14, 1931, two months after Winkler's successful launch of a similar device, the identical thrust chamber, embedded in an egg-shaped container filled with stagnant water, propelled a test arrangement of two parallel-mounted tanks which were pulled up by the propellant lines to an altitude of 60 feet. A few days later the only slightly damaged rocket was repaired and performed another flight to about 180 feet altitude. In the latter part of 1931, more flights were conducted with an improved rocket which was still pulled by the nose-mounted motor, but had the oxygen and fuel tanks arranged in tandem. The tail of the rocket accommodated a beer can-sized container with a parachute. There were several successful flights in which the pencil-shaped rocket rose to 1,000 or 1,500 feet and descended on the parachute so it could be used again.

These flights and their attendant publicity attracted the interest of the Ordnance Department of the 100,000-man Reichswehr, as the German Army was known then. Rocketry offered an approach to a long-range weapon unrestricted by the Treaty of Versailles whose authors had limited their provisions to classical artillery. One day in the spring of 1932, the Raketenflugplatz had three inconspicuous visitors, all in mufti. They were Colonel Doctor Karl Becker, chief of ballistics and ammunition, Major von Horstig, his ammunition expert, and Captain Doctor Walter Dornberger, in charge of solid rockets for the field artillery.

After a thorough briefing on the Raketenflugplatz's test and flight activities and its future plans, an agreement was signed under which, for a payment of 1,360 marks, an advanced and still untested version of VfR's rocket was to be test flown on the Reichswehr artillery range of Kummersdorf, sixty miles south of Berlin. This site was equipped with elaborate photographic equipment to track the rocket's flight path.

The launching took place on a cloudless day in July 1932 and was only partly successful. The rocket rose from its guide rails and soared upward. But by the time an altitude of some 3,300 feet had been reached, the trajectory had become almost horizontal. The rocket crashed into the trees about two miles from the launch site before the parachute could open.

Discussions with Colonel Becker after the flight test ended in a deadlock. The Ordnance Department said it was interested in the potential of the liquid-fuel rocket as a long-range weapon, but felt the Raketenflugplatz should produce sound test-stand performance data on rocket motors rather than firing "toy rockets." Also, it should stay away from showmanship and publicity which were incompatible with sensitive Reichswehr objectives along these lines.

Nebel argued that there was no money for elaborate test-stand instrumentation and that launchings and showmanship were vitally necessary to raise the funds to operate the Raketenflugplatz. He did not prevail. Becker made it clear that there would be no military support for the activities at Reinickendorf. However, he was prepared to support serious liquid rocket development work as long as it was conducted behind the fence of an Army enclave. This offer Nebel declined.

Becker had a deep-seated concern that any public mention of the Ordnance Department's support of liquid-propellant rocket development might bring about an international controversy which could only result in stopping the Reichswehr from exploring this unique opportunity to circumvent legally the long-range artillery limitations imposed by the Versailles Treaty. This concern was not limited to Nebel and his operation. Other independent rocketeers and space writers such as Winkler, Engel, and Ley maintained a rambling correspondence with rocket and space-flight enthusiasts all over the world. Becker felt strongly that this had to stop if the Reichswehr was to go into a hush-hush development of long-range rockets. Just as the word "atom," for centuries an international scientific term without any military connotation, fell under a military veil when a few years later the United States initiated its "Manhattan Project," Becker wanted to remove the word "Rakete" from the German dictionary.

Those rocketeers who accepted Army offers and contracts had to abide by the new strict classification rules. Among those who failed to receive such contracts or who refused to submit to the new secrecy, there was, of course, some grumbling. Some of them said later they suspected the Army had used its secrecy clout mainly in order to eliminate the competition by the "independents" for its own greater glory. However, not one of the "independents" had either successful rockets or the resources to develop them independently.

In the fall of 1932 Wernher von Braun, who had assisted Klaus Riedel in his designs with theoretical calculations while pursuing his engineering studies at the Berlin Institute of Technology, graduated from that school. When this came to Colonel Becker's attention, he invited von Braun to accept a research grant funded by the Ordnance Department, under which he could experimentally and theoretically investigate the combustion processes in a liquid-propellant rocket motor. The experimental work would be conducted on an existing test stand at the Kummersdorf artillery range. Becker would arrange for the study report to be accepted as a doctoral thesis by the University of Berlin, where he also held a full professorship. The report itself would be classified "Secret." Von Braun accepted this attractive invitation, although it meant that he had to sever his association with the Raketenflugplatz. He was extremely skeptical about the future of the Reinickendorf establishment, anyway. To him, the Repulsor route was a trifling approach to a real liquid fuel rocket. When he thought of all the new elements needed to make a real rocket work, gyro controls, jet vanes, actuators, cut-off control, feed pumps, and electromagnetic valves, he was sure that Reinickendorf was utterly inadequate even to commence such a vast program. It seemed that the funds and facilities of the Army were the only practical approach available to advance the cause of space flight. Adolf Hitler, of course, was not yet in power and did not figure at all in this reasoning. Nobody at that time, and that included great statesmen in other countries, predicted the magnitude of his future aggressive exploits. The moral issue, therefore, of the acceptance of an Army research grant in rocketry at that time did not differ from that faced by any of the many aviation pioneers accepting grants or contracts from their respective armed forces.

For two more years, however, Reinickendorf did keep going. On June 29, 1933, an unsuccessful launch was attempted with a sizable rocket of 260 pounds take-off weight. Its four parallel tanks were pulled by the nose-mounted motor. On August 11, 1933, the same rocket was launched again, this time from a raft in Schwielow Lake near Berlin. Because of a faulty valve, it rose to an altitude of only 270 feet and crashed into the lake with the motor still giving off irregular puffs of power.

The actual dissolution of the Raketenflugplatz in 1934 had a tragicomical aspect. It was unable to pay its water bill. The free lease had not included water, and several taps had been dripping away for years in abandoned buildings. When the water bill was presented, it was truly astronomical. It was the end of the Raketenflugplatz.

When on November 1, 1932, von Braun began work for his doctoral thesis on a remote test site of the Kummersdorf Proving Ground, he did not imagine that a mere five years later most of his friends at the Raketenflugplatz would be working together with him again at the great rocketry center at Peenemünde. For the moment, however, his operating base was tiny even by Reinickendorf standards. His laboratory was one half a concrete pit with a sliding roof, the other half being devoted to solid-fuel rocket work. His staff consisted of one mechanic, Heinrich Grünow, and his work orders were often overlooked in an artillery shop filled with jobs of higher priority than his.

Nevertheless, in January 1933, a first, small water-cooled alcohol-liquid-oxygen motor was ready for static test. To the amazement of Dornberger, von Braun's new boss, it developed a thrust of 310 pounds for 60 seconds at its first test. Half a year later, a 660-pound motor, regeneratively cooled by its own alcohol flow, passed a series of tests with flying colors. The time had come to build a flyable rocket.

Six months later the rocket, A-1, was ready, and a fraction of a second after the button had been pushed it was a shambles. Delayed ignition detonated an explosive mixture which had accumulated in the combustion chamber.

A second A-1 with improved ignition might have flown well, but other factors called for a complete redesign named A-2. In this, the large flywheel which had been in the nose of A-1 was moved to a location between the propellant tanks near the center of gravity. As with a rifle bullet the brute force of this gyro was to effect stabilization. A-2 had no vanes or rudders.

A few days before Christmas 1934, two A-2s (christened "Max" and "Moritz") were launched from the island of Borkum in the North Sea. Both rose vertically to an altitude of 1½ miles and were judged complete successes. A few months earlier, his thesis on injection, combustion, and expansion phenomena in liquid-propellant rocket motors had won von Braun a Ph.D. in physics.

His Kummersdorf work force had also grown from its two-man nucleus. During the summer of 1934, Walter Riedel, who had continued the late Max Valier's experimental work with Army Ordnance support, joined the team. A few months later came Arthur Rudolph, another of Valier's co-workers. In his spare time, Rudolph had built a complete alcohol-oxygen rocket power plant of pure copper which made two wholly successful static runs at first trial.

The successful flights of Max and Moritz gave quite a spurt to the Kummersdorf team, and soon a highly sophisticated A-3 was on the drawing boards. It was 22 feet long and was to be powered by a 3,200-pound-thrust alcohol-oxygen rocket engine that would operate for 45 seconds; and it was to have a full-fledged, three-dimensional gyro control system with jet rudders and rudder actuators. A-3's mission, besides demonstrating all these advances in technology, was to lift a considerable load of recording instruments to the upper atmosphere and to return safely by parachute. Originally, the design called for supersonic velocities, but continued addition of supplementary loads increased the weight excessively. Among the innovations piled into A-3 was a liquid-nitrogen pressurization system with an electrically heated vaporizer, in lieu of the heavy high-pressure flask of A-2. Novel alcohol and oxygen valves were operated pneumatically by magnetic servo valves. A two-stage flow feature was incorporated to eliminate the hazard of ignition explosions.

In December 1937 three A-3 launch attempts were made, and all three failed. The A-3 was the first rocket to be launched without guide rails. On December 4 the first ignited perfectly and rose beautifully from its launch platform. It continued for about five seconds until the parachute emerged, streaming into the fiery jet which consumed it in an instant. The bird began to spin, tumble, and crashed into the sea.

In the hope that the parachute and its mechanism had been responsible for the failure, it was omitted when the second A-3 was launched, but without success. Again the rocket spun, tumbled, and crashed.

By now it was clear that there was something wrong with the rocket's gyroscopic control system. It had been successfully tested with the entire rocket suspended in gimbals (three-axis suspension) while being fired up on the static test stand. But on the test stand there had been no wind effects. It seemed that the strong sea breeze during the launch induced a roll movement which caused gimbal lock and subsequent tumbling of the gyroscopic platform. Some doubt arose whether the jet rudders were large enough and whether they were moving fast enough to suppress the roll movement.

It was decided to wait for a windless day for the test of the remaining A-3, to be launched without alterations. Alas, number three malfunctioned as had the others.

A complete redesign of the control system seemed to be necessary. Since it would take eighteen months for the new system to be ready for flight tests, the time was used to introduce several other changes in the A-3. The fuselage was redesigned. A newfangled four-channel telemeter transmitter was installed to radio in light data to the ground; and the shape of the tail fins was changed in accordance with the latest data from the new supersonic wind tunnel at Peenemünde (see below). The modified A-3 would not have a supersonic capability, but it was to serve as a scale model for a much bigger A-4 rocket that was to come thereafter. The modified A-3 was to be known as A-5, in order to avoid the stigma of the earlier failures.

In the summer of 1939, shortly before the outbreak of World War II, the first flight test with the A-5 and its new guidance system proved that the lesson learned from the A-3 failures had been well heeded. It was a full success. During the next two years some twenty-five A-5s were launched, some of them even twice after successful parachute landings and recovery from the water.

In early 1935, more than two years before the abortive A-3 flights, the German Luftwaffe (Air Force) had placed a development contract with von Braun's fledgling Kummersdorf test station for an alcohol-liquid-oxygen powerplant of 2,200 pounds thrust which was to be mounted into a propeller-driven Heinkel 112 fighter airplane. The first successful static tests of the system, installed in a jacked-up fuselage and operated from the cockpit, were carried out during the summer of 1935. They were observed by a group of Luftwaffe officers who were first incredulous and then amazed.

The Luftwaffe was eager to give the little Kummersdorf establishment additional tasks to exploit further the rocket potential for aviation. They wanted the von Braun group to redesign the pressure-fed rocket system of the Heinkel 112 into a pump-fed one suitable for incorporation into a truly futuristic all-rocket fighter tentatively labeled Heinkel 176. In addition, they wanted a pair of alcohol-oxygen rocket-assisted take-off devices for heavy bombers that could be parachuted back to the airfield for repeated use. For the Luftwaffe to let such substantial development contracts to an Army artillery test station was as unusual in the Germany of 1935 as it would be in the United States today, but interservice rivalry and etiquette were quickly forgotten under the impact of the noisy, fiery test demonstrations.

However, the Kummersdorf station, sandwiched between two

General Walter Dornberger and Dr. Wernher von Braun at Peenemünde.

artillery ranges, was too small to accommodate all of these demands in addition to the development of the Army's own A-3 rocket. When the Luftwaffe offered 5 million reichsmarks to establish more elaborate facilities in another location, the Army topped this bid with another 6 million to retain controlling influence. For Dornberger and von Braun, whose Kummersdorf budget had never exceeded 80,000 reichsmarks per year, this opened a new world of possibilities. (One reichsmark equaled about 25 U.S. cents officially.)

In early 1936 the first construction hut was erected in the forests near the Baltic fishing village of Peenemünde, on the island of Usedom, where the elaborate new joint Army-Luftwaffe facility was to be built. In April 1937 the Kummersdorf group moved in and seemed lost in the tremendous plant. Klaus Riedel, Hans Hüter, Kurt Heinisch, and Helmuth Zoike of the now-defunct Raketenflugplatz joined the new, enlarged team.

The Peenemünde establishment consisted of a Luftwaffe-operated airfield for the testing of rocket-powered aircraft and an Army-operated research and development plant for guided ballistic missiles. Von Braun was appointed technical director of the Army complex, whose military commander, Lieutenant Colonel Leo Zanssen, reported to Walter Dornberger, by now a full colonel and department head for rocket development, both liquid and solid, at the Ordnance Department in Berlin.

After the A-3 failure in 1937, Dornberger found that his Army superiors made their continuing support of the liquid rocket program contingent upon its more direct usefulness for military applications. A-5, the A-3's successor, was only capable of a range a little in excess of ten miles and could not carry any payload except two parachutes and a telemetry transmitter. "We cannot hope to stay in business," said Dornberger to his Peenemünde engineers, "if we continue indefinitely to fire experimental rockets. The Ordnance Department wants a field weapon which can carry a sizable warhead over a range substantially exceeding that of long-range artillery."

Computations showed that a rocket of a configuration similar to A-5 but scaled up to the maximum dimensions that could be transported through railway tunnels, could cover a range of 170 miles carrying a warhead of one metric ton, or 2,200 pounds. The rocket would be about 46 feet long, have a diameter of 5.2 feet and a launch weight of approximately 26,000 pounds. All this satisfied the higher echelons, and in this informal way, in late 1937, the original concept of the A-4—in later years to become known as V-2—came into being.

It was decided that the motor of the A-4 would be pump-fed. Various types of pumps, even including piston pumps, were discussed with potential manufacturers. Pump and turbine test equipment was installed at Peenemünde and a development contract was let for an experimental centrifugal pump for liquid oxygen. In the summer of 1940 a flyable turbo-driven pump to feed the alcohol and oxygen into the A-4 motor was ready for integration into the rocket.

The turbine drive was the next problem. The exhaust gas from the prime alcohol-oxygen combustion was clearly too hot for the turbine blades. One alternate method was to inject water into an alcohol-oxygen flame produced within a small separate combustion chamber or gas generator. Another was to reduce the gas temperature by running this gas generator at a very alcohol-rich mixture ratio. Both schemes, however, would have required a complicated maze of tubing, valves, and governors and presented a tricky start-up problem, as the feed pressures for the gas generators were zero before the turbopump started to spin. The development of a gas generator running on the A-4's main propellants was continued as a side-line effort and led to a limited laboratory success by 1944. However, the A-4 design itself was committed to the simpler solution of driving the turbine with the steam generated by the catalytic decomposition of hydrogen peroxide.

In 1934 Helmut Walter in Kiel had begun, under auspices of the German Navy, to develop a system that would permit a turbine-powered submarine to run submerged at high speed. His idea was basically the following. During a surface or a half-submerged run using a retractable air intake known as a "snorkel," fuel oil would be burned in pumped-in atmospheric air. The flame would heat a steam boiler, with the steam then driving the turbine and the propeller shaft. Under water, a stream of high-percentage hydrogen peroxide, to be carried in a separate tank, would be injected into a catalyst chamber where it would mix with potassium permanganate injected through another nozzle. The hydrogen peroxide, H_2O_2, would decompose into a hot, oxygen-enriched water steam according to the formula

$$2\ H_2O_2 \rightarrow 2\ H_2O + O_2 + \text{heat}$$

Into this steam, fuel oil would be injected to burn in the oxygen, and water would be added to reduce the steam temperature to a value acceptable to the turbine blades.

Wherever simplicity or light weight was more important than fuel economy, it was also possible to use only the first portion of this cycle, i.e., the hydrogen peroxide decomposition, and dispense with the secondary fuel oil injection and combustion. This concept, known as the "cold Walter cycle" in contrast to the "hot Walter" used in the subs, was offered by Walter to the Luftwaffe as a simpler (while less energetic) power source for rocket-assisted take-off devices and even for all-rocket aircraft.

It seemed a natural for Peenemünde to ask Walter to build a steam generator for the A-4's turbine-powered alcohol-oxygen pump on this principle. In 1941 the first peroxide steam generator was ready for installation.

Then there was the task of developing the powerful 55,000-pound-thrust rocket motor for the A-4. In 1937 Dr. Walter Thiel had taken over Wernher von Braun's old establishment at Kummersdorf to study and perfect combustion chambers. While reducing their size enormously, he succeeded in increasing combustion efficiency to better than 95 per cent. At Kummersdorf, however, the test stands had not permitted his working on motors of more than 8,000 pounds' thrust, and most of his tests were done at the A-5 thrust level of 3,200 pounds.

Thiel's investigations showed that it required hundreds of test runs to tune the propellant injection system of a rocket motor to maximum performance, and the prospect of having to do this with the A-4's 55,000-pound-thrust motor was frightening. So several small, well-proven injection systems, each installed in its now cup-shaped pre-chamber, were discharged into a single, large combustion chamber. The first test with the "18-cup motor" was an agreeable surprise, in that it showed higher efficiency than a single injection unit discharging into a small motor. Thus the 18-cup motor was never abandoned despite its cumbrousness and complications. Not until shortly before the end of the war was this motor's performance duplicated by a simpler design.

To answer the many unprecedented questions relating to the A-4's flight at speeds up to five times the speed of sound, a large supersonic wind tunnel was erected at Peenemünde under the direction of Dr. Rudolf Hermann.

Guidance and control of the A-4 posed further problems. There were competent industrial firms available that knew how to build gyroscopes or autopilots for aircraft, but they needed detailed specifications for the A-4's exotic flight conditions. These could only be developed at Peenemünde with the help of elaborate analogue computers and electronic simulators. From this need sprang Peenemünde's elaborate guidance and control laboratory under Dr. Ernst Steinhoff. The mathematical genius behind the emerging art of supersonic flight control was Dr. Hermann Steuding. In addition, Peenemünde contracted with a sizable number of university institutes for general scientific consultation and the exploratory development in fields such as integrating accelerometers, trajectory tracking by Doppler radio, research on radio wave propagation, missile antenna patterns, new measuring methods for the supersonic wind tunnel, analogue computers, and so forth.

The first attempt to launch an A-4 took place in the spring of 1942. The 12-ton missile rose from its launch platform for about one second until the fuel-feed malfunctioned and allowed it to settle back upon its fins. These lacked the strength for a hard landing with the result that the missile toppled over and disintegrated in a great explosion.

Four weeks later, the second A-4 promised to meet the most optimistic expectations. It passed through the dreaded sonic barrier without incident, but at the forty-fifth flight second, it began to oscillate. A cloud of white steam emerged and the missile broke apart in mid-air.

With a reinforced instrument compartment between the dummy warhead and the alcohol tank, the third V-2 was readied for launch and on October 3, 1942, it opened a new era for rocketry. After a flawless launch, it accelerated for sixty-three seconds, steady as a rock, to a top speed of 4,400 feet per second, at which time the rocket motor was cut off. It reached a maximum altitude of 52 miles and a range of 116 miles. For the first time, a man-made object had left the atmosphere and reached airless outer space.

The rest of the A-4 story was less spectacular and certainly less pleasant, not only for those who were to sit at the receiving end of its operational use against London or Antwerp, but also for its originators. After the first successful flight there were again many failures. Later, while most of the birds would successfully pass through the powered flight phase, it turned out that a goodly portion disintegrated during re-entry in the atmosphere.

Soon, there was also competition. Not to be outdone in the long-range missile field by the German Army, the Luftwaffe had embarked upon the development of a winged, subsonic cruise missile powered by an airbreathing pulse-jet propulsion system. This missile later on saw extensive operational deployment against the city of London and was labeled V-1 by the Propaganda Ministry. Compared to the A-4 (or V-2), its vulnerability to flak artillery, fighter aircraft, and even barrages of tethered balloons was partly compensated by its much lower cost.

Meanwhile, the war came to Peenemünde. On August 17, 1943, nearly 600 four-engine bombers of the Royal Air Force, escorted by 45 scouts and night-fighters, raided the Peenemünde installations and left almost 800 casualties. This raid, along with three others against factories near Berlin and Vienna, as well as one

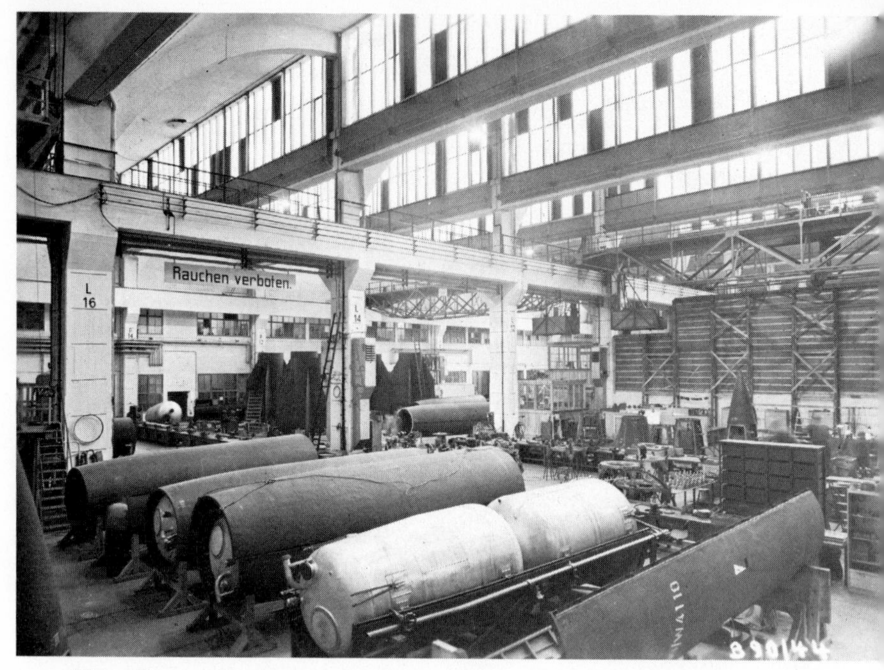

V-2 pilot production at the Army Research Center at Peenemünde, 1943.

against the famed Zeppelin Works at Friedrichshafen, all of which had been earmarked as production assembly plants for A-4s, caused Hitler to order the whole rocket production underground. The implementation of this order was turned over to an SS general, which led to evergrowing friction between the Army and the SS.

The deployment of the missile by military units caused many additional problems. Some were related to turning a highly sophisticated piece of machinery over to a group of men with limited technical background. Others had to do with the rapidly deteriorating strategic position of the German Army in the field.

The Peenemünde engineering staff had insisted from the outset that a weapon as new-fangled and complex as an A-4 rocket could not be fired successfully in combat except from elaborate concrete installations containing repair and testing facilities. By early 1944, however, after Hitler had decided to deploy the A-4 against London, Allied air superiority over the French Channel coast had become so overwhelming that Dornberger's earlier predictions that such installations would be bombed out of existence before they could go into operations were fully borne out. Dornberger, now a major general, wanted A-4 operations in the form of mobile batteries. He persisted on this course and was surprisingly successful. The great rockets would be trucked on specially designed chassis to sites usually located in dense forests. They would then be raised upright on their fins and fueled by special tank trucks while, with the help of a single rope the tops of adjacent trees would be pulled together thus forming canopies above the rocket's tips. A few seconds before a rocket's motor was ignited from an armored car, the canopy rope was released, the trees would elastically swing apart and the rocket could clear the treetops. This simple camouflage scheme was so successful that not a single mobile A-4 battery was ever bombed.

The A-4 was put into action against London on September 7, 1944, without any announcement to the German public. When the announcement was finally made, the A-4 was dubbed V-2, or "Retaliation Weapon 2," by Dr. Joseph Goebbels' propaganda writers.

Three other Peenemünde developments are worth recording: rocket-propelled aircraft, the Wasserfall guided antiaircraft missile, and the winged A-9.

A series of successful flights with the first rocket-powered Heinkel 112, piloted by Erich Warsitz in 1937, led to the development of an all-rocket fighter Heinkel 176, a very small aircraft with a pressurized, detachable, and parachutable cockpit. It had "wet" wings that accommodated the fuel and many other futuristic features.

Warsitz made several successful take-offs and landings with a He-176 powered by an interim "cold Walter" rocket; but, with the outbreak of World War II, the project was shelved. It was only toward the end of the war that rocket-powered interceptors staged a dramatic comeback, with Alexander Lippisch's tailless Messerschmitt 163.

In the winter of 1942–43, the growing Allied air threat caused the German Antiaircraft Command to ask Peenemünde to develop a radio-guided antiaircraft rocket. After a thorough defense systems study, in which an attempt was made to optimize

As World War II drew to a close, Allied scientific and technical intelligence experts made a sustained effort to intercept the Peenemünde rocket development team and capture as much V-2 documentation and hardware as possible. The Americans were particularly successful in this endeavor and, shortly after the war, began launching V-2s from the White Sands Missile Range in New Mexico. This rocket, adapted to carry upper atmospheric research instruments, is being made ready for flight in 1947. (U. S. Army)

the size and range of the missile (less range meant that more batteries would be needed to defend a given site), a configuration 26 feet in length was selected. It looked like a small A-4, with two pairs of crossed wings and enlarged air vanes permitting increased maneuverability in the atmosphere. The missile's fuel was a mixture of hydrocarbons while nitric acid was used as the oxidizer. The program director was Ludwig Roth, the former head of Peenemünde's advanced planning group.

Wasserfall, as the missile was named, made forty-four experimental flights of which the majority were successful. As the war ended, it could be maneuvered from the ground with a control stick, and a fully automatic hookup to an analogue computer (which developed steering commands from the ever-changing looking angles of a target-tracking and a missile-tracking radar) was partially completed. A proximity fuse designed to explode Wasserfall's warhead at the point closest to the target was to assure that even a near-miss would bring the airplane down.

The A-9 was an A-4 with wings. Two were flight tested during the winter of 1944–45. While the first one fizzled, the second performed a successful near-vertical climb to four times the speed of sound.

The objective of A-9 was to double the range of the A-4 by utilizing some of the tremendous kinetic energy available after cut-off for an extended aerodynamic glide. After the A-4 launch sites in France and Holland had been lost as a result of the Allied invasion of Normandy, there was much military interest in an extension of the range, even at the price of greater vulnerability to defense measures due to the A-9's slower target approach speed.

But drawings in Peenemünde's advanced planning section included those of a manned A-9, complete with pressurized cockpit and tricycle landing gear. Theoretically, it could carry its pilot over a distance of 400 miles in 17 minutes. When placed on a 200-ton thrust booster, A-9 could carry its pilot across the Atlantic, and with an additional still more powerful boost stage, the A-9 could attain orbit.

It was with such hopes for the future in mind that Wernher von Braun and the cream of his Peenemünde staff came to the United States when the war was over.

Men working on the inside of the propellant tank of an Atlas ICBM. (General Dynamics)

8

THE MILITARY BALANCE

In October 1945 the United States Army Air Force invited selected aircraft companies to submit proposals for participation in a long-range guided missile development program. The potential contractors were free to propose winged, air breathing, "cruise" missiles, or ballistic rockets with ranges from 20 miles to a maximum of 6,000 miles.

One of the successful bidders was Convair (Consolidated Vultee Aircraft Corporation), whose engineers suggested picking up where the Germans had left off with their V-2. Convair's first rocket, called the MX-774, was purely experimental and not designed to carry a military warhead. It was 34 feet long, 30 inches in diameter, burned liquid oxygen and alcohol, and at first glance looked very much like a scaled-down V-2. Upon closer scrutiny, however, its design philosophy showed marked differences. For one, the rocket's outer shell, made of aluminum alloy, served also as the wall of the oxygen and alcohol tanks, whereas the V-2, which was built to survive supersonic re-entry, had a separate and rather heavy steel hull which penalized its performance. Convair reasoned that since their MX-774 was only a precursor for a future ballistic rocket of several thousand miles' range, which would require warhead separation prior to re-entry anyway, the MX-774 would not need a rocket body capable of sustaining re-entry.

Another major difference was the propulsion arrangement. Whereas the V-2 was propelled by a single 55,000-pound thrust rocket engine, MX-774 had four separate, parallel-mounted motors of 2,000 pounds' thrust each. These motors, developed by Reaction Motors, Inc., were individually suspended in gimbal mounts, which permitted their swiveling for flight-path control. Thus, the V-2's drag-producing graphite steering vanes could be omitted.

In July, September, and December of 1948, three MX-774s were launched at the Army's newly established White Sands Proving Ground in New Mexico. All three lifted off their launch pads as expected, but unfortunately their motors cut off prematurely during climb. The fault was traced to a component (a solenoid valve) that was found to have failed under severe vibration. Nevertheless, many important lessons were learned; and when, in 1951, the Air Force, now an independent branch of the Armed Services, instituted the Atlas Intercontinental Ballistic Missile (ICBM) program, Convair received the assignment.

The Air Force's 1951 decision to go at once for intercontinental range was a bit like a child's resolution to learn to walk before he could crawl. However, it was not the result of arbitrary and reckless planning, but of dire necessity. The wartime alliance with the Soviet Union had given way to confrontations and Cold War, and there was plenty of disquieting intelligence that Stalin had directed the establishment of a crash program for the development of an atom-tipped rocket missile capable of flying over the North Pole from the Soviet Union to targets in the United States. As we shall see, he succeeded with this program and really beat the United States to the punch.

Atlas, the hoped-for answer to the Soviet long-range rocket challenge, was by no means the first United States design for a missile with intercontinental range. Right after the end of World War II, Northrop Aircraft Inc. received a contract from the then Army Air Force to develop the Snark, a swept-wing, turbojet-powered missile capable of carrying a heavy warhead over a range of 5,000 miles. After a solid rocket-boosted launch from a ramp, Snark would climb to an altitude of 50,000 feet and subsequently cruise at a velocity of 550 miles per hour, or Mach 0.9, just below the speed of sound. At this speed, about equal to that of a modern jetliner, it would need a little over eight hours to reach its distant target in the Soviet Union.

Built for the U. S. Air Force by North American Aviation (now Rockwell International), the Navaho XSM-64 was canceled in July 1957 after ten years of development. The rocket-boosted, ramjet-sustained missile, however, left a rich technical legacy, which was soon applied to large liquid-propellant rocket engines, aerodynamics, thermodynamics, guidance, and material processing. The missile was 95 feet long, weighed 305,000 pounds, and was boosted by a 415,000-pound-thrust three-chamber rocket engine. Eleven Navahos were fired between 1956 and 1958. One reached a distance of nearly 2,000 miles in 1957. (Rockwell International)

After many trials and tribulations—some people caustically referred to the Atlantic Ocean off Cape Canaveral as "Snark-infested waters"—the winged missile went into limited production as an unmanned supplement to the new Strategic Air Command's emerging fleet of B-52 bombers. Air Force planners, of course, were fully aware that Snarks were vulnerable not only to conventional flak but to antiaircraft guided missiles that even then were on the drawing boards. However, since Snark's design was based on conventional technology and was thus a weapon "in hand," development and production continued. Every Snark attacked by antiaircraft weapons, so the reasoning went, would detract fire from a manned B-52.

There was also the far more ambitious Air Force Navaho project, contracted to North American Aviation. An intermediate-range test version of about 3,000 miles range was called the SM-64, while the full-scale weapons system of 5,000 miles range was known as the SM-64A. Both versions of Navaho were powered by supersonic ramjet engines (engines using rammed-in air as the oxidizer) capable of sustained cruise at three times the speed of sound, or about 2,000 mph. The ramjet-powered vehicle would be boosted to its cruise speed and altitude by riding piggyback on a large booster powered by two liquid-propellant rocket engines.

Three successful launches with the interim version proved the soundness of the Navaho concept. The booster engine drove the huge twin missile through the sound barrier to Mach 3, the ramjet engine ignited, the booster dropped off, and the cruise missile was on its way. But just as Navaho's designers saw the fruits of their many years of labor, the Air Force abruptly withdrew its support and canceled the project.

What was behind this change of heart? Ever since a caveman raised a wooden plank to ward off a rock hurled at him by an enemy emerging from the bush, weapons development has been a continuous race between defensive and offensive capabilities. During World War II German missilemen heatedly argued the pros and cons of winged, air-breathing, cruise-type missiles and wingless ballistic rockets. The former concept had been embraced in Germany by the Luftwaffe and led to the pulse-jet powered V-1; the latter resulted in the German Army's V-2.

Now, less than ten years later, the same argument flared up between American missilemen. Using aviation's classical yardstick of ton-mile transportation economy, the advocates of the cruise missile found it easy to show that a ton of military payload (the warhead) could be carried over a given range more economically, and with less take-off weight, by using air-breathing engines. The advocates of the ballistic missile would counter that the warhead of an intercontinental rocket would approach its target at a speed of 15,000 mph, as contrasted to a Navaho's approaching at 2,000 mph and even a Snark at the relative snail's pace of 550 mph. Defensive weapons, they would argue, were once again on the upsurge, and by the time Snark and Navaho could be deployed operationally, they would be sitting ducks for those novel guided surface-to-air antiaircraft missiles (SAM) that America's Cold War adversary, the Soviet Union, was also known to be developing.

However, the issue was not quite that straightforward. In 1950 no one had yet demonstrated convincingly that a guidance system could be built that would enable a ballistic rocket to hit a target 5,000 or 6,000 miles away with any acceptable accuracy. The task was difficult enough for a cruise missile, but electronic and inertial guidance systems for air-breathers could at least be adequately tested in long-range aircraft, whereas this was not possible with equipment designed for the very different—and far less benign—environment of a ballistic rocket.

A breakthrough in warhead development ultimately tilted the balance in favor of the ballistic rocket. An ICBM with its expensive propulsion and guidance system had to be "cost effective," i.e., the damage wrought in the target area had to be substantially greater than the cost of the missile. With a TNT-tipped warhead such as the V-2 had, cost effectiveness was simply an unattainable goal for missiles of intercontinental range, because the more TNT one placed into the nose, the larger, heavier, and costlier the rocket became. Even with a nuclear fission bomb of the Hiroshima type as a warhead, the cost effectiveness of an ICBM was at best marginal. But when the spectacular 1952, 1953, and 1954 thermonuclear tests at Bikini and Eniwetok demonstrated the feasibility of a light-weight hydrogen fusion bomb, an ICBM of reasonable dimensions became a real possibility.

A Spartan missile, the long-range interceptor missile of the Safeguard ballistic missile defense system, is launched from Meck Island at the U. S. Army's Kwajalein Missile Range in the Pacific. Spartan can be launched both singly and in salvos and is guided by Safeguard missile site radar and its associated data processor to numerous successful intercepts of target ICBMs fired from California during developmental and system testing at the range. Spartan has also intercepted target IRBMs fired by the Navy from a surface test ship. (U. S. Army)

Moreover, the tremendous wallop its warhead would pack could still demolish any "soft" target even if the missile's guidance was not precise enough for a direct hit.

In the wake of this warhead breakthrough, the U. S. Air Force, in August 1954, stepped up the Atlas effort and established an extended crash program for the development of an arsenal of hydrogen-tipped ICBMs. Complete authority and control over all R & D and testing aspects of this program, whose cost was ultimately to exceed the famed wartime Manhattan Project, was assigned to a newly-formed organization with the nondescript title of "Western Development Division." It was located in Los Angeles and commanded by the young and able Major General Bernard A. Schriever. Colonel Otto Glasser became Schriever's project officer for the Atlas, which was contracted, as we have seen, to Convair. The engines for Atlas were to be developed by North American Aviation in view of their success with the Navaho booster engines. Overall systems engineering was contracted to the Ramo-Wooldridge Corporation which had assembled a galaxy of talent in all the new technologies involved.

The United States decision to seriously get going on a massive ICBM strike capability had by now been long overdue, for in 1954 there were ominous signs that the Soviet Union's determined long-range missile program, initiated years earlier, was beginning to bear visible fruit.

Long-range strike capability was a novelty in the Russian military set-up, which traditionally had stressed massive army ground forces ever since the Muscovite dukes freed their country from the yoke of the Mongolian Golden Horde. For a vast, landlocked country this emphasis on ground forces was only natural. Even in World War II, when the Soviet Union was in mortal danger and American and British long-range bomber forces were pounding the industrial heartland of Germany, there was little evidence of a comparable Soviet strategic air force.

In the postwar years Joseph Stalin and the men around him fully realized that henceforth the emphasis would have to be changed. The only power capable of stopping the further spread of Soviet hegemony would be the United States, but ground forces alone would be pretty useless in any confrontation with a world power on the other side of the globe.

For the Soviet Union to build up a strategic air force in a few years would have been quite difficult. True, Russia had excellent aerodynamicists and aircraft designers, but both America and Britain were years ahead in the field of long-range bombers. Moreover, the build-up of a highly sophisticated Soviet aircraft industry would have to take place at a time when the country was still digging out from wartime rubble. For this very reason, the development of long-range, air-breathing cruise missiles such as the American Snark or Navaho was not a very attractive option for the Soviet Union either.

How about rockets?

The Red Army had made massive use of salvos of solid-propellant rockets (launched from the famous "Stalin organs") against the invading German Armed Forces, but these were short-range weapons for tactical support of ground forces. And now, in the late 1940s, the underground production assembly facility of German V-2 rockets near Nordhausen, south of the Harz Mountains, had been taken over by the Red Army. Although the U. S. Army had evacuated about a hundred V-2s in various stages of completion and shipped them to America, there were still many parts left in the production pipeline and supplier plants in East Germany.

The first act for the Kremlin was therefore to decide to assemble several hundred more V-2s from these parts and ship them to the Soviet Union for evaluation. But while much useful technology could be syphoned from such a program, the V-2 with its puny 200-mile range was clearly not the final answer Stalin was seeking.

Skimming through German documents, Soviet intelligence had come across a paper study by Eugen Sänger, describing a hypersonic, rocket-powered bomber which, skipping through the upper layers of the atmosphere, could deliver a bomb to a target halfway around the world. Stalin, who had always taken a great personal interest in rocketry, demanded a thorough briefing and got so enthusiastic about the potential of Sänger's antipodal bomber that he dispatched a team (including his son Vasilii) to Germany to bring Sänger to the Soviet Union. Sänger, however, had wound up in French custody after the armistice, so the project came to naught.

The plan to assemble and evaluate V-2s was nonetheless carried out. Several hundred of them were shipped to the Soviet Union and launched for the familiarization of military crews with the problems of handling complex guided missiles. Others were used for the scientific exploration of the upper atmosphere, and still others were analyzed by propulsion engineers or guidance experts working on more advanced rocket designs. A few years later, a re-engineered V-2, with an improved Russian rocket motor known as the M-101 and a range of about 400 miles, was placed in production.

However, the often repeated story that the Soviet missile program (which in a relatively short time led to the world's first intercontinental ballistic missile and soon thereafter to Sputnik, the pioneering artificial satellite) had been the work of German rocket engineers taken into the Soviet Union after the war, is just plain nonsense. Wernher von Braun's development team, which had fathered the V-2, had left its Peenemünde rocket center on the Baltic before the Red Army arrived and was ultimately captured by the United States Army in southern Germany. The Soviet Army did draft a number of production engineers from the subterranean plant in the Harz Mountains back into service after the war to get the V-2 production moving again. Most of these men were indeed moved to the Soviet Union later, but, being production rather than development people, they could contribute little to more advanced missile concepts. There were a few first-class development engineers among them, but according to their own accounts (published years later), they were merely asked to offer their own ideas for possible technological improvements without being given access to secret Soviet plans for the future. Under these conditions, the work of the former Peenemünde men in the Soviet Union was mainly in support of the evaluation and improvement of the V-2. By 1953 all of them had been repatriated to Germany.

The assignment to develop a Soviet ICBM was given to Chief Marshal of Aviation N. F. Zhigarev and his special weapons chief, General A. S. Yakolov. The Z.I.A.M. research institute and the Khimki facility, both in or near Moscow, were designated as key program centers, while most of the rocket engine development work went to the Frunze engine plant (named for Mikhail V. Frunze, a revolutionary hero) in Kuibyshev. Launchings of V-2s, surface-to-air missiles, and rockets of moderate range were conducted at Kapustin Yar near the point where the Volga River flows into the Caspian Sea. When the big ICBMs were ready for launch tests, they were taken to the newly established "Kennedy Space Center of the Soviet Union," the missile and spaceport of Tyura Tam, east of the Aral Sea, in the Kazakh Soviet Socialist Republic. From there, the intercontinental rockets arced on a northeasterly heading toward distant targets in the Pacific Ocean.

Undoubtedly, the managers of the Soviet ICBM program were confronted with the same payload/range/accuracy problem that plagued their American counterparts. But there was one important difference. America emerged from World War II with a powerful Strategic Air Command capable of delivering relatively light-weight atomic bombs. The Soviet Union had no comparable long-range air arm and was still trying frantically to explode its first superheavy atomic "device." When, in the late forties, Soviet missilemen asked their atomic bomb colleagues for realistic weight estimates for nuclear warheads that could be made available for ICBM use within the next few years, the weight figures based on their relative lack of sophistication must have been colossal compared to those quoted in the United States. Heavy payloads, however, meant enormous rockets. It does not take much imagination to guess Stalin's probable reaction to the appalling size of the rockets his missilemen told him would be required to lug a Soviet atomic warhead to the United States: "I don't care how big it is. Build it!"

This decision not only gave the Soviets the first—albeit clumsy—ICBM. It also enabled them in 1957 to launch Sputnik and beat the United States in putting the first man in space. (Unlike nuclear warheads, men cannot be miniaturized, so a smaller rocket would not be adequate.)

During the latter part of World War II, the U. S. Army had carefully evaluated all intelligence reports on the V-2 rocket and knew what opportunities and what individuals to look for once the hostilities had ended. The man in charge of the round-up

Production model of the Redstone missile at the final check-out station, at the Michigan Ordnance Missile Plant, Warren, Michigan. (Chrysler Corporation)

operation was Colonel Holger N. Toftoy, U. S. Army Ordnance. His first priority action was to evacuate the priceless hardware booty from the underground Harz Mountain production assembly plant before, according to the Yalta agreements, the area in which it was located would be turned over to the Red Army. Next, Toftoy established a special interrogation camp in Garmisch-Partenkirchen in Bavaria, where Dornberger, von Braun and several hundred former Peenemünde men were interviewed by teams of American experts. These contacts ultimately led to "Project Paperclip," so named because the dossiers of Germans handpicked during these interviews were marked with a paperclip. Under Paperclip, 120 former V-2 engineers including Wernher von Braun were brought to the United States between late 1945 and the summer of 1946.

During their first three years in America, the Paperclip group assisted in the launching of V-2s at White Sands both for technological evaluation and high altitude research. The group was also instrumental in the two-stage Bumper-Wac project wherein a Wac Corporal rocket of American origin was launched from the nose of a V-2. In 1949, one Bumper-Wac reached the unprecedented altitude of nearly 250 miles. But the Paperclip group was unable to bring about radical advances in rocket technology because of lack of funds and suitable ground test facilities. Under the gentle tutorship of U.S. Army Major James P. Hamill, the Germans, now housed at Fort Bliss, Texas, conducted some research on a supersonic ramjet engine designed to propel a cruise missile launched by a V-2, but even that modest effort progressed slowly because of the utterly limited resources.

In 1950, Toftoy, Hamill's boss and by now a major general, succeeded in having the Redstone Arsenal in Huntsville, Alabama, assigned to him as the Army's future guided-missile arsenal. Hamill's Paperclip group, meanwhile augmented by over 400 Americans, moved from Texas to Huntsville and became the nucleus of the new arsenal staff. Its first assignment was the development of the Redstone, a ballistic rocket capable of carrying a rather heavy thermonuclear warhead with great accuracy to a target 200 miles away. (The Redstone made its first partly successful flight on August 20, 1953.)

While a nuclear-tipped rocket with 200 miles' range was of great significance for the field support of the Army ground forces, it was of course no answer to the strategic challenge posed by the Soviet push for an intercontinental rocket. Response to this strategic challenge was assigned, as we have already seen, to the U. S. Air Force's Strategic Air Command and its mixed aircraft-missile retaliatory force.

In late 1955 the signs of rapid Soviet progress with their ICBM became more ominous and the political clouds darkened. This was the time of the Berlin airlift. The Atlas and General Schriever's "second-generation" ICBM, the 9,000-mile Titan, were still in the early testing phase and years away from opera-

The 69-foot-long Redstone 200-mile-range ballistic missile being made ready for launch at the Cape Canaveral launch site in late 1956. Often called the child of the V-2, development got under way in 1951 and the first firing took place on August 20, 1953. (U. S. Army)

tional readiness. The first Atlas was successfully fired on December 17, 1957, but it fell short when it impacted only about 500 miles away. Only on November 28, 1958, did it reach its full intercontinental range of 6,325 miles. Faced with this situation, the Pentagon decided to initiate an additional crash program aimed at the development of not one but two intermediate range ballistic missiles (IRBM), the Army's Jupiter and the Air Force's Thor. Each was to carry a lightweight thermonuclear warhead over a range of 1,500 miles. The Thor was assigned to General Schriever's Western Development Division in Los Angeles, while responsibility for Jupiter, along with the now-operational Redstone, was given to the energetic and resourceful Major General John B. Medaris, commander of the newly created Army Ballistic Missile Agency in Huntsville, Alabama. Wernher von Braun and his team, by now American citizens, became Medaris' technical arm.

The basic idea behind the IRBMs was as obvious as it was controversial. It was obvious because a ballistic missile of only 1,500-mile range was closer to the general state of the art of missilery and easier to guide accurately than its big intercontinental brother. It was controversial because an IRBM would have to be stationed on the territory of some of America's North Atlantic Treaty Organization (NATO) allies. Yet any IRBM launch site would automatically become a prime target for Soviet missiles. Moreover, logistics support of IRBM bases would depend on ships vulnerable to Soviet submarines. Finally, non-American NATO launch crews would get uncomfortably close to the sensitive trigger of the United States nuclear deterrent capability. But for the moment, there was no other answer to the growing Soviet threat.

In the latter part of 1958 Thors were deployed in England and Jupiters in Italy and Turkey. The Pentagon had assigned over-all operational control for both missiles to the U. S. Air Force in accordance with its global strategic mission. Though the Army was not happy about this decision, the alleged waste caused by interservice bickering was grossly overplayed by the news media.

That the services did indeed work very closely together was best attested to by the co-operation between the Army and the

An Atlas 19E ICBM erected in Launch Complex 567–9A at Fairchild Air Force Base in Washington, July 1961. (General Dynamics/U. S. Air Force)

U. S. Navy on Jupiter. Because of the IRBM's inherent dependence on allied launch sites, the Navy decided to study the possibility of launching Jupiters from converted Liberty ships. This task involved the carrying or even manufacturing of substantial quantities of liquid oxygen aboard ship. Also, in order to protect herself effectively against enemy submarines, the Liberty ship had to retain her freedom of movement and unrestricted maneuverability while preparing for the launching of her missiles, possibly in dense fog or at night. This requirement could only be met by continuous updating of the Jupiter's guidance system from the position, heading, and speed readouts of the ship's own inertial navigation system.

The "sea-borne Jupiter" program was ultimately canceled, but the joint Army-Navy effort to solve this tricky guidance problem laid the scientific basis for an aiming system used to this day with the Navy's Polaris, Poseidon, and Trident missiles which are launched from nuclear-powered submarines. Under the leadership of Rear Admiral William F. Raborn, Jr., and his successor, Rear Admiral Levering Smith, these Navy missiles have over the years become a cornerstone of the United States nuclear deterrent force. Unlike land-based IRBMs and ICBMs, missile-carrying submarines, constantly moving and hidden in the depths of the vast oceans, are virtually invulnerable.

In the mid-sixties, the Atlas and Titan ICBMs had proven their operational capability and the Jupiters and Thors were gradually withdrawn. In the latter sixties, the original Titan, which used liquid oxygen as oxidizer, began to be replaced by the newer Titan 2, whose tanks were loaded with noncryogenic, storable liquid propellants. Nevertheless, even with room-temperature propellants, it was quite a chore to keep a missile in a status of fifteen-minute alert readiness over a period of several years. As the continuous efforts to further miniaturize both warheads and guidance systems met with success, the Air Force initiated the development of the all-solid fuel, multistage Minuteman ICBM, which was ultimately to replace the Atlas and the Titan 2 as well.

For a while, the "balance of terror" between the two superpowers was determined by the numerical count of ICBMs on

America's first IRBM, the Jupiter, was developed at the Army Ballistic Missile Agency and manufactured by the Chrysler Corporation. Jupiter's first full-range flight occurred on May 31, 1957. Here it is seen just before take-off at Cape Canaveral, Florida. (U. S. Army)

both sides. The world seemed to relax a bit in the shadow of the enormous annihilating capability accumulated in the two opposing arsenals. Clearly, the destructive power had become so immense that in any nuclear holocaust, both sides would perish. Many were reminded of that famous animal behavior experiment which showed that two male scorpions locked in a bottle will not harm each other, although in the open they are known to engage in combat ending invariably in the death of one. In the confinement of the bottle, neither scorpion will attack the other since each seems to be aware of the fact that even if it struck first, there would be enough fight left in the trapped, dying opponent to retaliate with a deadly and inescapable sting.

However, soon there were signs of possible destabilizing effects in the uneasy balance of terror. Suppose one side "hardened" (missile jargon for strengthening against near-misses of atomic bombs) its missile site while the other did not? Or suppose only one party acquired an antimissile system that could reliably protect its missile sites as well as its cities and industrial complexes?

When the Cold War showed signs of thawing in the late sixties, the United States and the Soviet Union decided to talk things over. Both sides readily agreed that an unchecked missile race was not only fraught with enormous dangers, but also drained the economies of both countries. But from an agreement on a basic truth that was so obvious that it almost became a platitude, to an ironclad and cheat-proof arms control agreement was—and is—a difficult and rocky road. Delicate questions such as mutual inspections of top secret missile sites arose. What good was an agreement on missile numbers if both sides were free to increase the "bang" delivered by the warheads or to place several warheads into the nose of a single rocket? Suppose that these multiple warheads could even be independently targeted and were maneuverable to evade antimissile missiles? If one of the two sides decided not to keep up this endless race, would it become a "sick scorpion," possibly tempting its bottle mate to use the opportunity to strike for fear that it may one day become sick and an opportune victim of the other?

This is the eerie backdrop of the SALT (strategic arms limitations talks) meetings, possibly the most fateful meetings in which man was ever involved. For the moment it looks as if the huge rockets with their awesome hydrogen warheads are keeping the big powers at peace in spite of their many disagreements and conflicting interests. Only the future can tell whether, as we all hope and pray, the age of the great world wars will be a thing of the past—forever.

Installed above the Mariner 9 is a single rocket engine, used to make course corrections en route to Mars and to brake velocity as the spacecraft approaches and enters into orbit around the red planet. This 300-pound-thrust engine operated on monoethyl hydrazine and nitrogen tetroxide propellants, which made up some 45 per cent of the weight of the entire spacecraft. (NASA)

9

THE QUEST FOR NEW WORLDS

Man began dreaming of flights to the Moon and to other heavenly targets long before the most rudimentary of rockets made their appearance. Yet man's notion of those other worlds was nebulous; he did not know of the limited height of our atmosphere and that beyond there was only airless space. Bird power was therefore the preferred propulsion mode of his early dream voyages to other worlds. Balloons and even guns were also pressed into service, and whenever a writer of early space science fiction did not feel inclined to bother with the engineering details of his tale, he simply invoked magic powers or a mysterious little black box that lifted and moved him at will and whose secrets only the inventor knew.

We have already seen how, during the first half of the twentieth century, the intellectual and technological base had been created to harness the power of large and efficient rockets for man's quest for new worlds. By 1956 the time had arrived for an action plan.

The science academies of sixty-seven nations had agreed to designate the two-year period from January 1, 1957, to December 31, 1958, as the International Geophysical Year (IGY), and to pool their national scientific resources in a co-ordinated data-gathering program aimed at studying "the Earth as a planet." The idea was to collect jointly facts on global phenomena such as the Earth's magnetic field and its fluctuations: aurora phenomena in the Arctic and Antarctic regions, atmospheric, wind, and climatic patterns; oceanic currents; and other aspects of man's physical environment. As part of their national contributions, both the United States and the Soviet Union announced their intent to launch geophysical research satellites into orbit. The United States promptly set about implementing that obligation, but few people in this country really had taken the Soviet declaration of intent very seriously. In the view of most American scientists and engineers involved in the IGY discussions, the general state of Soviet technology as evidenced in World War II could not possibly support a project as ambitious as an artificial satellite. The ominous warnings from the intelligence community that the Soviets were making rapid progress with the development of an intercontinental ballistic rocket were held under tight security wraps. Among scientists without access to this information, the Soviet announcement was therefore widely accepted as an act of wishful thinking on the part of a few dreamy-eyed Russian physicists.

On the American side it was clear that such a satellite, the first man-made object ever to be placed in orbit, should be extremely simple. Of course, in order to prove that orbit had been attained, the satellite had to be equipped with a tiny radio transmitter. As a scientific by-product, the radio signal, through accurate tracking of the orbital path, would make it possible to survey precisely the little irregularities and anomalies in Earth's gravitational field. In addition, it could also transmit to ground stations a limited quantity of data from space, such as spacecraft temperature in sunlight and Earth shadow, the number and time of micrometeorite hits, and the intensity of cosmic radiation. More ambitious space science projects would clearly have to wait until later.

The key question was, what kind of a rocket should be used to launch the first satellite? In 1956, a special committee was appointed to make selection from among three proposed options for a suitable orbital launch vehicle:

(1) An Air Force Atlas intercontinental ballistic rocket. At that time the Atlas had never flown. It made its first partly successful flight over a distance of 500 miles on December 17, 1957, but did not demonstrate its full intercontinental capability until November 28, 1958. In 1956, however, the Air Force still thought that Atlas could be readied in time for the IGY. With a

greatly reduced payload weight, it would be capable of attaining orbit.

(2) Juno 1, a slightly modified Army Redstone medium-range ballistic rocket with a three-stage cluster of small solid rockets placed in its nose. The basic Redstone itself had by then performed many successful flights over a 200-mile range. The modification would involve the replacement of ethyl alcohol by dimethyl-hydrazine, a more potent fuel that had been thoroughly static tested in the Redstone engine, and a moderate "stretch" in the length of the propellant tanks. The solid rockets suggested were well proven. The spin-up table, required to provide spin stabilization for the cluster of high-speed stages, would be straightforward engineering. In the eyes of several committee members, the whole Juno 1 configuration smacked of a jury-rigged rocket and clearly lacked sophistication.

(3) A brand-new Navy Vanguard rocket. The Vanguard proposal envisioned a highly advanced three-stage rocket. The first stage was to be based on the general technology used in the Navy's highly successful Viking research rocket, but all the rest of Vanguard was to be developed from scratch.

After long deliberations, the committee selected option 3. Since the IGY was a peaceful international science project, members felt it would be inappropriate to use military hardware such as the Air Force's Atlas or the Army's Redstone as launch vehicles. Unfortunately, the Soviet Union did not share such a lofty sentiment. It committed its repeatedly flight-tested 20-engine ICBM rocket to do the job. The result of this uneven contest was Sputnik 1's unforgettable "beep-beep" on October 4, 1957, and a lot of red faces in Washington. But partly in reaction to the Sputnik embarrassment, America buckled down to serious business and in July 1969—twelve years later—landed the first men on the Moon!

In the meantime, while faces were still red (and they got even redder as the first Vanguard was destroyed in a spectacular and widely televised explosion on the launch pad), the Army was hurriedly commissioned to launch its four-stage Juno 1, and eighty days—119 days after Sputnik 1—on January 31, 1958, America' first satellite, Explorer 1, was in orbit.

Now that the race was on, satellite after satellite followed and soon both the ill-fated Vanguard and the new-fangled Atlas proved that they, too, could "carry the goodies." During 1966 alone, the United States successfully launched no less than ninety-seven artificial satellites, space probes, and manned spacecraft. The ice was broken. The big rockets had opened up space for a wide new arena of human activity.

On March 3, 1959, from Cape Kennedy, the first man-made object was launched into a trajectory that carried it forever out of the gravitational field of the Earth. The spacecraft itself, Pioneer 4, had a minimum of scientific instrumentation but its telemeter transmitter provided ironclad proof that it had indeed zoomed past the Moon at a speed sufficient to escape Earth's gravity. The launch rocket, a Juno 2, was another make-do configuration from the Army's stable: the three-stage cluster of small solid rockets of the successful Juno 1 was simply transplanted from the Redstone into the nose of the more powerful Jupiter IRBM. Pioneer 4 blazed the trail for a whole fleet of highly sophisticated interplanetary spacecraft, including Mariners to Mars, Venus, and Mercury, and Pioneers to Jupiter and beyond.

On April 12, 1961, the world was stunned by a new and magnificent Soviet first. Cosmonaut Yuri Gagarin, launched by a rocket designated A-1 that had originally been designed as an ICBM, was orbiting the Earth in his 10,395-pound Vostok 1 spacecraft. All that the United States could offer in immediate response was a 200-mile suborbital ballistic ride down the Atlantic Ocean, on May 5, 1961, by Astronaut Alan Shepard atop a standard Redstone. On August 6, 1961, the Soviets topped Gagarin's two-orbit flight with a seventeen-orbit, twenty-five-hour flight by Cosmonaut Gherman Titov. Only on February 20, 1962, could Astronaut John Glenn finally demonstrate that America was capable of manned orbital flights, too. Encased in a Mercury capsule, an Atlas launched him into orbit, for three trips around the Earth.

With the manned launch of March 23, 1965, the single-seat Mercury flights had given way to the two-seat Gemini capsules. To accommodate their greater weight, an advanced second-genera-

Atlas-Mercury launch vehicles in final assembly dock. (General Dynamics)

tion ICBM, the Titan 2, was pressed into service. But again, the Soviets beat America to the punch: on October 12, 1964, the Soviets launched Voshkod 1, a three-man spacecraft, which orbited the Earth for twenty-four hours before being successfully recovered.

On October 11, 1968, for the first launch of an American three-seat Apollo spacecraft, the manned space program began to use launch rockets that had not been designed originally to carry military warheads. Apollo 7, the first manned Apollo flight, with Walter Schirra in command, was launched into orbit from Cape Kennedy by a Saturn 1B rocket. But the eight engines which powered its first stage had emerged from the military IRBM and ICBM programs and only the upper stage had no military progenitors. It was powered by a single engine that used exotic liquid hydrogen as fuel and liquid oxygen as oxidizer.

For a long time Apollo 7 was destined to be the only manned spacecraft launched by a Saturn 1B, for Apollo 8 and all subsequent Apollo missions would require a launch vehicle capable of hurling the spacecraft all the way to the Moon. The performance requirements of the huge Saturn 5 that ultimately boosted all of the American astronauts on their lunar missions were not, as one might expect, firmly established before actual construction began.

In 1961 President John F. Kennedy had committed the United States to "land a man on the Moon before this decade was out." The first decision the National Aeronautics and Space Administration (NASA) then had to make was how, exactly, the lunar flight mission was to be executed. Three conceivable flight profiles were studied in much detail:

(1) The direct mode, where a single, very powerful, launch rocket would inject into translunar trajectory a heavy spacecraft capable of directly soft-landing on and subsequently reascending from the Moon. It turned out that this mode would have required a first stage with eight instead of Saturn 5's five rocket engines.

(2) The Earth-orbit rendezvous mode. Here, the launch rocket's third stage, that must drive the same heavy spacecraft of Mode 1 from low Earth orbit to translunar injection, would be refueled in Earth orbit by propellants carried up by a second identical launch vehicle. It was found, that with this concept the total boost load could be distributed on two smaller five-engine Saturn 5s.

(3) The lunar-orbit rendezvous mode. Here, a single launch vehicle—it was hoped that a five-engine Saturn 5 would suffice—would launch a spacecraft configuration consisting of three modules directly into a translunar trajectory. The propulsion system of the service module would first induce itself and the two other modules into an orbit around the Moon. The command

The first stage of Saturn 1B AS-203 rocket before being hoisted into the vertical position for checkout and testing. (NASA)

module, in which the astronauts would ultimately return to Earth, would remain in lunar orbit, still attached to the service module, while a lunar module, solely designed for the lunar surface portion of the flight plan, would descend from lunar orbit to the Moon's surface and later on reascend to the orbit for a rendezvous with the command and service module. The latter two modules would return to Earth and only the heat-protected command module would survive the blazing re-entry into and through the atmosphere.

Mode 3 was ultimately selected. However, before that decision could be made, many detailed studies on how Earth–Moon flight could be accomplished had to be conducted. Also, it was not clear at first what total payload-carrying capability would be required of the Saturn 5 rocket in order to inject the three-module spacecraft into translunar trajectory. Two weight figures best illustrate the degree of uncertainty. When the first manned lunar mission was launched (Apollo 8, lunar orbit only, Christmas 1968, Frank Borman commanding) the weight injected into translunar trajectory was just under 87,400 pounds. By comparison, the last manned lunar mission, Apollo 17 (that visited the Taurus-Littrow region in December 1972), under the command of Eugene Cernan, weighed 108,000 pounds.

The steady weight growth resulted from the desire to increase the scientific usefulness of the Apollo missions. Whereas the first flights landed in smooth mare areas close to the lunar equator, later missions were dispatched into tricky mountainous regions at higher lunar latitudes. Also, a lunar surface vehicle called the Rover was added to the payload of Apollo 15, 16, and 17, to extend the astronauts' land excursions on the Moon. Apollo lunar scientific experiment packages (ALSEPs), powered by long-lasting radioactive thermoelectric generators, were taken along and deposited on several landing sites. Five ALSEPs formed a seismic network that for years kept radioing back information on lunar earthquakes and meteor impacts on the Moon's surface. In addition, astronaut stay times on the Moon were substantially extended. All of this meant not only a greater payload weight on the lunar surface, but also the requirement for more propellants in the lunar module's descent stage to de-

Working on the first stage of the mighty Saturn 5 three-stage launch vehicle. Each of the Rocketdyne F-1 engines develops more than 1,500,000 pounds of thrust. (NASA)

Apollo 16 astronauts Thomas K. Mattingly II, John W. Young, and Charles M. Duke, Jr., start their journey moonward the afternoon of April 16, 1972. On this mission, as on all others in the Apollo series, the Saturn 5 performed according to the book. (NASA)

orbit and soft-land that additional weight. For the Saturn 5 launch rocket, the payload requirements grew by leaps and bounds. Toward the end of the Apollo flights, the original performance padding for "unforeseen factors" was used up, and every conceivable trick to save weight and soup up engine performance was employed to squeeze more payload performance out of the mighty rocket.

After Apollo 17, Saturn 5 was used one more time, to carry Skylab, America's first space station, into Earth orbit. As the main laboratory and crew compartment of Skylab had been fashioned from a third stage of a Saturn 5, the over-all launch configuration did not differ greatly from that of the Saturn/Apollo flights. And since injection into a low Earth orbit required much less terminal velocity than a flight to the Moon, the two lower stages of Saturn 5 were entirely adequate to launch the 100-ton, 118-foot-long Skylab.

Skylab was occupied for periods of 28, 59, and 84 days by three separate three-man crews of astronaut-scientists. These crews rode up into Earth orbit in Apollo command modules boosted by Saturn 1B rockets.

Television coverage of Apollo launches (and, to a lesser degree, of Skylab launches) gave the Saturns much public exposure. Meanwhile, however, and with much less fanfare, another launch vehicle became increasingly popular as a carrier for various unmanned spacecraft. Titan 3, made up of a modified Titan 2 ICBM to which two solid booster rockets had been side-strapped, had originally been envisioned by the U. S. Air Force as a launch vehicle for their now defunct manned orbiting laboratory. When this program—a small military orbital station—was canceled, Titan 3 with its favorable price tag and its great flexibility for carrying payloads to low as well as synchronous (twenty-four-hour) orbits, attracted more and more traffic.

The Atlas ICBM, to which an Agena top stage had been added, likewise became a popular space-launch vehicle. Later, the storable-propellant Agena stage was often replaced by the more powerful liquid hydrogen-burning Centaur stage, which technologically had helped pave the way for the Saturn hydro-

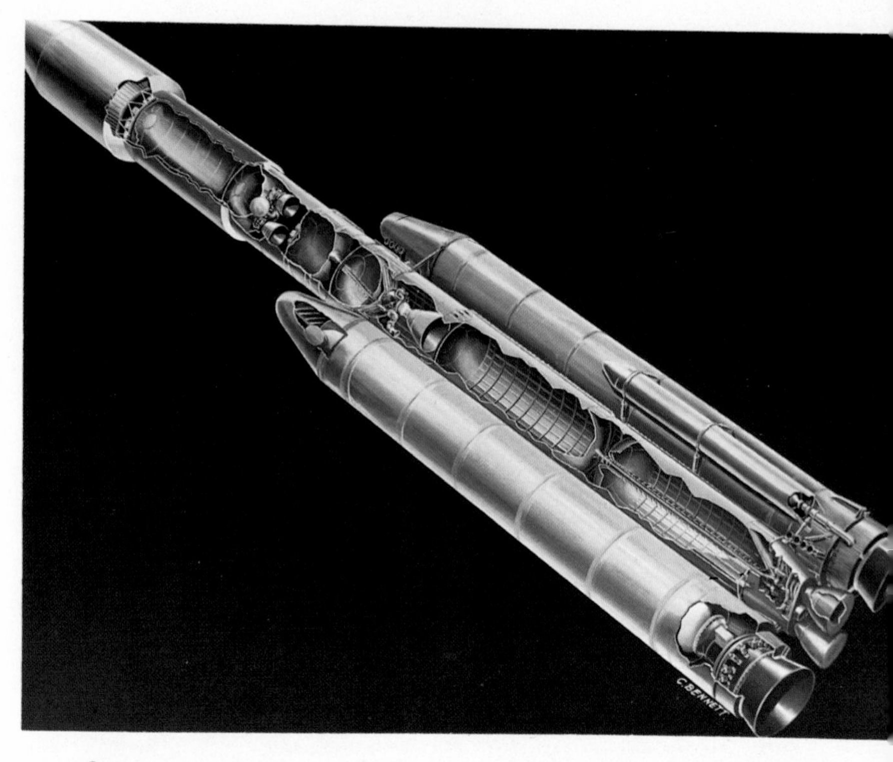

Cutaway view of the Titan 3D with the Centaur upper stage. Sometimes referred to as Titan 3E, its two solid-propellant booster rockets develop 2,400,000 pounds of thrust, the lower liquid-propellant stage 520,000 pounds, the second liquid stage 101,000 pounds, and finally the Centaur 30,000 pounds. This version of the Titan weighs 1,416,000 pounds at launch and can place 34,000 pounds of payload in a low Earth orbit. (Martin Marietta Aerospace)

gen-powered upper stages. An Atlas-Centaur launch vehicle injected the fabulously successful Mariner 9 spacecraft into an orbit around Mars from which its television cameras sent reams of unprecedented surface pictures back to Earth. Atlas-Centaurs also served as launch vehicles for the Jupiter fly-by Pioneers 10 and 11, the fastest spacecraft ever dispatched from the Earth. Recently, the Centaur has also found its way into the nose of a Titan 3.

In addition to these high-powered, heavyweight rockets, the unfolding space age had an evergrowing need for less expensive launch vehicles with a moderate speed and payload capability. The lowest end of the performance spectrum was covered by the four-stage, all solid Scout rocket. But most of the space traffic between the Scout and the heavyweights was captured by a variety of multistage versions of the U. S. Air Force Thor, originally designed as a single-stage 1,500-mile IRBM. The sales catalogue of Thor varieties attempted to offer individual payload customers such as NASA or Intelsat (the international communications satellite consortium) or ESRO (the European Space Research Organization) the lowest possible launch price and highest demonstrated reliability by applying the principle of maximum commonality of hardware. In other words, each commercial or scientific customer could benefit from the collective experience of many previous flights with a similar launch vehicle of the Thor family, while he did not have to pay for unnecessary lifting power in case he had a lighter payload or a lesser injection-speed requirement. The spectrum of Thor-based launch vehicle configurations included Thor-Able, Thor-Agena, Thor-Delta, solid-propellant thrust augmented Thors with all of these upper stages, and finally some of these combinations with an additional solid-fuel top stage for higher terminal velocities.

In aeronautical terms, any type of space flight involves very high speeds. Even to attain a low orbit, the terminal velocity of the launch rocket must be on the order of Mach 24. The propellant requirement to orbit a pound of payload is therefore quite high even for the most economical launch rockets. However, once the payload has been injected into orbit, it will stay there unpowered for many years. In terms of "miles-per-gallon," a satellite can therefore be quite economical. If it stays in orbit long enough, it will ultimately beat a Honda scooter for fuel economy. Nevertheless, the launch cost is still a major portion of the price of any commercial or scientific satellite, and any reduction of the launch cost is bound to extend the usefulness of space operations for the benefit of man.

This is where NASA's latest project, the reusable Space Shuttle, comes in. Regardless of whether we use a small Thor-Delta, a medium-sized Titan 3, or a giant Saturn 5, it still costs about $500.00 per pound net payload to place an object into a low orbit, and about $1,000.00 per pound if we want to recover that object. The Shuttle, due to its reusability, will carry 65,000 pounds of payload to orbit and back for about $10 million per flight, which comes out to be $154 per pound.

There will be other advantages. The Shuttle will open up space for nonastronauts. Anyone healthy enough to ride an airliner as a passenger should be able to ride the Shuttle to orbit. Astronomers, meteorologists, Earth resources prospectors, medical researchers, materials processing engineers, to name a few, will have access to the myriad opportunities offered through observations from orbit or by the unique zero-gravity environment in space. In addition, the list of future Shuttle riders may well include test technicians who will be able to give the unmanned satellite (manufactured by their company and carried aloft in the same Shuttle flight) a final in-orbit checkout before cutting its navel cord and setting it free up there. One of the most heartbreaking aspects of unmanned space launches is "spacecraft infant mortality." Upon successful injection into orbit by a conventional launch rocket an unmanned spacecraft must be taken through the various phases of orbital activation: its power-generating solar panels must be deployed, its antennas extended, its attitude stabilization and thermal control systems activated, its camera covers opened, its telemetry transmitter turned on, and so forth. A single mishap at any stage of this critical activation sequence will convert a shiny multimillion-dollar satellite into a hapless piece of orbiting junk.

NASA launched an unmanned spacecraft from Cape Kennedy on April 8, 1966, atop an Atlas-Agena launch vehicle. The Orbiting Astronomical Observatory payload was the first in a series of four designed to give astronomers their first sustained look into the universe from above the obscuring and distorting effects of the Earth's atmosphere. (NASA)

Thor-Delta No. 101 used in a Westar 1 communications satellite mission, April 13, 1974. Take-off thrust was augmented by nine solid rocket boosters. (NASA)

The Shuttle offers an attractive way out of this dilemma. In the future, operational users may buy some of their spacecraft "f.o.b. orbit." The Shuttle, being not only the most economical space launch vehicle around, but also capable of taking passengers along, will accommodate factory representatives who can take their satellite through those critical deployment phases while, via manipulator arms, it is still firmly connected to the Shuttle. Should any difficulties arise during the spacecraft's activation, the technicians would simply pull the satellite back into the Shuttle's cargo bay, close the hatch, return to Earth, and have the factory fix the trouble.

In due time the Shuttle, because of its inherent cost-effectiveness, will replace our entire stable of throwaway launch vehicles. It will also end the old argument as to whether the United States should or should not abandon manned space flight altogether and limit itself to the less expensive exploration and utilization of outer space with unmanned equipment. The Shuttle's reusability will make it an attractive carrier for continued manned space activities, as well as the most economical truck to carry unmanned spacecraft to orbit. There will always be a need for many unmanned communications, meteorological or scientific satellites; but if they, too, can be orbited at the lowest possible cost with the help of a manned, reusable orbital carrier, the cost argument about manned versus unmanned space flight will simply evaporate.

The more than eight hundred experiments and observations conducted by the nine Skylab astronauts have rendered impressive and lasting proof of the value of man's presence in space as an observer, operator, updater, communicator and even repairman. The Shuttle will thus give us the best of both worlds: unmanned spacecraft for highly economical, long-life missions such as communications satellites where man's presence in orbit can be of little assistance; and elaborate orbital observatories, laboratories and even manufacturing facilities whose potential could never be fully developed without the presence of man.

FROM THE COLLECTION OF FREDERICK I. ORDWAY

Frederick I. Ordway III and Wernher von Braun

Wernher von Braun, father of our space program, was director of NASA's George C. Marshall Space Flight Center in Huntsville, Alabama. He is presently vice president of Fairchild Industries in Germantown, Maryland, and president of the National Space Institute, a non-profit educational organization in Washington, D.C.

Frederick I. Ordway III started his career in rocketry at America's pioneering rocket engine firm, Reaction Motors, Inc., and later joined the von Braun team at the Army Ballistic Missile Agency and NASA's Marshall Center. More recently he became a professor at the University of Alabama in Huntsville, School of Graduate Programs and Research. Among his many books is the widely acclaimed *History of Rocketry and Space Travel*, co-authored with Dr. von Braun.

BIBLIOGRAPHY

Relatively few modern books have been written about the history of rocketry, and none claim to approach completeness. For this reason, a large part of the research undertaken for Chapters 1 through 6 has necessarily been based on materials written before the twentieth century. Books on the history of gunpowder and artillery often include rockets, and even if the attention paid to rocketry is scant, they permit one to appreciate the climate in which rocketry evolved. Such books include the following:

Bertholot, Marcelin, *La Chimie au Moyen Age* (three volumes). Paris: Imprimerie Nationale, 1893.

Bethell, Colonel H. A., *Use of Rockets in Artillery in the Field.* London: Macmillan, 1911.

Braun, Wernher von, and Frederick I. Ordway III, *History of Rocketry and Space Travel.* New York: Crowell, 1966, 1969, 1975.

Brock, Alan St. H., *A History of Fireworks.* London: Harrap, 1949.

———, *Pyrothecnics: The History and Art of Firework Making.* London: O'Connor, 1922.

Canby, Courtlandt, *A History of Rockets and Space.* New York: Hawthorn, 1963.

Clark, John D., *Ignition! An Informal History of Rocket Propellants.* New Brunswick, N.J.: Rutgers University Press, 1972.

Dickson, Katherine M., *History of Aeronautics and Astronautics—A Preliminary Bibliography.* Washington, D.C.: NASA, 1968.

Durant, Frederick C., III, and George S. James, eds., *First Steps Toward Space.* Washington D.C.: Smithsonian Institution Press, 1974.

Faber, Henry B., *Military Pyrotechnics.* 3 vols. Washington: Government Printing Office, 1919.

Guttmann, Oscar, *Monumenta Pulveris Pyrii.* Text in English, French, and German. London: Artists Press, 1906.

Held, Robert, *The Age of Firearms.* New York: Harper, 1957.

Hime, Lieutenant Colonel Henry W. L., *Gunpowder and Ammunition: Their Origin and Progress.* London: Longmans, Green, 1904.

———, *The Origin of Artillery.* London: Longmans, Green, 1915.

Hoeffer, Ferdinand, *Histoire de la Chimie Depuis les Temps les Plus Reculées Jusqu'à Notre Époque.* 2 vols. Paris: Fortin, Masson, 1842.

Hogg, O. F. G., *The Royal Arsenal: Its Background, Origin and Subsequent History.* London: Oxford, 1963.

Jocelyn, Colonel Julian R. J., *The History of the Royal Artillery.* London: Murray, 1911.

Ley, Willy, *Rockets, Missiles and Men in Space.* New York: Viking, 1968.

Magne, Émile, *Les Fêtes en Europe au XVIIe Siècle.* Paris: Martin-Dupuis, 1930.

Mourey, Gabriel, *Le Livre des Fêtes Françaises.* Paris: Librairie de France, 1930.

Partington, J. R., *A History of Greek Fire and Gunpowder.* Cambridge, Eng.: Heffer, 1960.

Pollard, Major A. B. C., *A History of Firearms.* London: Geoffrey Bles, 1926.

Reinaud, Joseph Toussaint, and Ildephonse Favé, *Histoire de l'Artillerie: Feu Grégeois, des Feux de Guerre et des Origines de la Poudre à Cannon.* 2 vols. Paris: Dumaine, 1845.

Riling, Ray, *Guns and Shooting: A Selected Chronological Bibliography.* New York: Greenberg, 1953.

Romocki, S. J. von, *Geschichte des Explosivstoffe.* Vol. 1, Berlin: Robert Oppenheim, 1895. Vol. 2, Hannover: Gebrüder Jänecke, 1896.

Schmidt, Rodolphe, *Développement des Armes à Feu.* Schaffhausen: Brodtmann, 1870.

Warner, Philip, *The Medieval Castle.* New York: Taplinger, 1971.

Wilkinson, Henry, *Engines of War.* London: Longman, Orme, Brown, Green & Longmans, 1841.

During the last several hundred years, many books have appeared on the use of rockets for both military and recreational purposes. The titles that follow relate to material appearing in Chapters 1, 2, 4, 5, and 6. No attempt is made to include journal and other documentation; this is a task presently under way in a two-volume work, *Rocketry Through the Ages* by Frederick I. Ordway III, Mitchell R. Sharpe, and Frank H. Winter, to be published by the University of Alabama Press in Tuscaloosa.

Alberti, G. A., *La Pirotechnia, o sia Trattado die Fuochi d'Artificio.* Venice: Recurti, 1749.

Anderson, Robert, *The Making of Rockets.* London: Morden, 1696.

Antonj, Domenico, *Trattato Teórico-Pratico.* Trieste: Sambo, 1893.

L'Arte de Fare i Fuochi d'Artifizio con poca Spesa. Napoli: Tasso, 1834.

Babington, John, *Pyrotechnia, or, A Discourse of Artificiall Fire-works.* London: Harper & Mab, 1635.

Bate, John, *The Mysteries of Nature and Art.* London: Mabb, 1635.

Ben, Joseph, *Erfahrungen über Congrev'schen Brand-Raketan . . .* Weimar: Landes-Industrie Comptoirs, 1820.

Biringuccio, Vanoccio, *De la Pirotechnia.* New York: American Institute of Mining and Metallurgical Engineers reprint, 1942 (originally appeared in 1540).

Blümel, Johann Daniel, *Luft-Feuerwerkerey.* Strasbourg: König, 1771.

Boillot, Joseph, *Modelles Artifices de Feu et Divers Instrumens de Guerre.* Chaumont-en-Bassigny: Mareschal, 1598.

Browne, William H., *The Art of Pyrotechny.* London: "The Bazaar" Office, 1883.

Chertier, François-Marie, *Nouvelles Recherches sur les Feux d'Artifice.* Paris: Chertier, 1854.

Colliado, Luigi, *Prattica Manuale dell'Artiglieria . . .* Milano: Chisolfi, 1641.

Congreve, William, *A Concise Account of the Origin and Progress of the Rocket System.* London: Whiting, 1807; Dublin: O'Neil, 1817.

———, *The Details of the Rocket System . . .* London: Whiting, 1814. (Reprinted in 1970 by the Museum Restoration Service, Ottawa, Ontario)

———, *The Different Modes of Use and Exercises of Rockets.* London: Whiting, 1808.

———, *Memoir on the Possibility, the Means and the Importance of the Destruction of the Boulogne Flotilla in the Present Crisis . . .* London: Whiting, 1806.

———, *A Treatise on the General Principles, Powers, and Facility of Application of the Congreve Rocket System . . .* London: Longman, Rees, Orme, Brown and Green, 1827.

Corréard, Joseph, *Histoire des Fusées de Guerre.* Paris: Corréard, 1841.

Cutbush, James, *A System of Pyrotechny.* Philadelphia: C. F. Cutbush, 1825.

Denisse, Amédée, *Traité Pratique Complet des Feux d'Artifice.* Paris: Denisse, 1882.

Dennett, John, *A Concise Description of a Powerful Species of War Rockets.* London: Dennett, 1832.

D.M., *Pyrotechnia of Konstige Vuurwerken* . . . Rotterdam: Ryckhals, 1672.

Dollecsek, Anton, *Geschichte der Österreichischen Artillerie.* Vienna: Kreisel & Gröger, 1887.

Ellena, Giuseppe, *Corso di Materiale d'Artiglieria.* Turin: Scuola d'Applicazione delle Armi d'Artiglieria e Genio, 1877.

Ferré Vallvé, Juan Bautista, *La Pirotecnia Moderna.* Barcelona: Soler, 1904.

Die Feuerwerkerei. Leipzig; Paul, n.d.

Frezier, Amédée François, *Traité des Feux d'Artifice.* Paris: Nyon, 1747; rev. ed., Paris: Jombert, 1747.

Furttenbach, Joseph, *Architectura Navalis.* Frankfurt-on-Main: Clemens Sleichen, 1629.

[Grignon], *La Pyrotechnie Pratique.* Paris: Cellot & Jombert, 1780.

Hale, William, *A Treatise on the Comparative Merits of a Rifle Gun and Rotary Rocket.* London: Mitchell, 1863.

Hassenstein, W., ed. and trans., *Das Feuerwerkbuch von 1420.* Munich: Deutsche Technik, 1941.

[Jombert, Charles-Antoine], ed., *Manuel de l'Artificier.* Paris: Jombert, 1757.

Jones, Captain Robert, *A New Treatise on Artificial Fireworks.* London: Millar, 1765.

———, *Artificial Fireworks.* Chelmsford: Meggy & Chalk, 1801.

Kentish, Thomas, *The Pyrotechnists' Treasury.* London: Chatto & Windus, 1878.

———, *The Complete Art of Firework-Making.* London: Chatto & Windus, 1905.

Kostantinov, Major General Konstantin I., *On Fighting Rockets.* St. Petersburg: Tipographia Eduard Veimar, 1864. (Principal text in French, title page in Russian.)

L——e. L. von, *Vollständiges Taschenbuch für Kunst und Lustfeuerwerker* . . . Budapest: Hartlebens, 1820.

[Leurechon, Jean (or Henry van Etten)], *Mathematicall Recreations.* London: Hawkins, 1633.
Lorrain, Hanzelet [Jean Appier], *La Pyrotechnie.* Pont-à-Mousson: Gaspard, 1630.
———, and François Thybourel, *Recueil de Plusieurs Machines Militaires, et Feux Artificiels pour la Guerre, & Recréation.* Pont-à-Mousson: Marchant, 1620.
Livre de Cannonerie et Artifice de Feu. Paris: Sertenas, 1561.
Majendie, Sir Vivian Dering, *Ammunition.* London: Mitchell, 1867.
Malthe, François de, *Traité des Feux Artificiel pour la Guerre, et pour la Recréation.* Paris: Guillemot, 1632.
Malthus, François, *Pratique de la Guerre.* Paris: Clovsier, 1650.
Malthus, Francis, *Treatise of Artificiall Fireworks Both for Warres and Recreation.* London: Hawkins, 1629.
Meyer, Franz Sales, *Die Feuerwerkerei.* Leipzig: Seemann, 1898.
Meyer, Moritz, *Traité de Pyrotechnie.* Liège: Oudart, 1844.
Mongéry, Merignon de, *Des Fusées de Guerre, Maintenant Fusées à la Congreve.* Paris: Bachelier, 1825.
Moore, William, *A Treatise on the Motion of Rockets.* London: Robinson, 1813.
Morel, A. M. Th., *Traité Pratique des Feux d'Artifice et pour la Guerre.* Paris: Didot, 1800.
Mortimer, G. W., *A Manual of Pyrotechny.* London: Simpkin & Marshall, 1824.
Nye, Nathaneal, *The Art of Gunnery.* London: Leak, 1647. (Includes "A Treatise of Artificiall Fireworks for Warre and Recreation.")
[d'Orval, Perrinet], *Essay sur les Feux d'Artifice pour le Spectacle et pour la Guerre.* Paris: Coustelier, 1745.
———, *Traité des Feux d'Artifice Pour le Spectacle et pour la Guerre.* Bern: Wagner & Muller, 1750.
Ossorio, Marcello Calà, *Instituzioni di Pirotecnia per Istruzione di Coloro che Vogliono Apprendere a Lavorare i Fuochi d'Artifizio.* Naples: Stamperia Reale, 1819.
Porta, John Baptista, *Natural Magick.* London: Young & Speed, 1658.
Pralon, A., *Les Fusées de Guerre en France.* Paris: Berger-Levrault, 1883.
Pyrotechny, or, The Art of Making Fireworks at Little Cost and with Complete Safety and Cleanliness. London: Ward, Lock and Tyler, 1873.
Rogier, Charles, *A Word for My King and Country: A Treatise on the*

Utility of a Rocket Armament . . . Macclesfield: Wilson, 1818.
Ruggieri, Claude-Fortuné, *Elemens de Pyrotechnie*. Paris: Barba & Magimel, 1802.
———, *Pyrotechnie Militaire*. Paris: Patris, 1812.
Ruggieri, Gaetano, and Giuseppi Sarti, *A Description of the Machine for the Fireworks* . . . London: Bowyer, 1749.
Scoffern, John, *Projectile Weapons of War and Explosive Compounds*. London: Cook & Whitley, 1852.
Simienowicz, Casimir, *The Great Art of Artillery*. London: Tonson, 1729. (English edition of Kazimierz Siemienowicz, *Artis Magnae Artilleriae*.)
Smith, Thomas, *The Art of Gunnery*. London 1643, to which is added "Certain Additions to the Book of Gunnery with a Supply of Fireworks."
Sonzogno, Cesare, *L'Arte di Fare i Fouchi d'Artifizio* . . . Milan: Sonzogno, 1819.
Susane, M., *Les Fusées de Guerre*. Metz: Blanc, 1865.
Thompson, C. R., *The Biography of Wm. Schermuly and the History of the Schermuly Pistol Rocket Apparatus Ltd.* London: Victoria House, 1946.
Todericiu, Doru, *Preistoria Rachetei Moderne Manuscrisul de la Sibiu (1400–1569)*. Bucharest: Editura Academiei, 1969.
Vegetius, *Feuerwerkbuch, 1420*. Augsburg: Steiner, 1529.
Venn, Thomas, *The Compleat Gunner*. London: Pawlet, 1672.
Vergnaud, M., *Manuel de l'Artificier*. Paris: Roret, 1826.

As for Chapter 3, "The Rocket in Asia," source material is not concentrated in readily available Western language books but is scattered quite widely. Chinese gunpowder and rocketry are covered in:

Davis, Tenney L., and James R. Ware, "Early Chinese Military Pyrotechnics," *Journal of Chemical Education*, 24 (Nov. 1947), 522–37.
Goodrich, L. C., and Fêng Chia-Shêng, "The Early Development of Firearms in China," *Isis*, 36, No. 104 (Pt. 2) (1946), 114–23, and "Addendum," 36, Nos. 105–6 (Pts. 3–4) 250–51.
Schlege, Gustave, "On the Invention and Use of Fire-Arms and Gunpowder in China," *T'oung Pao*, 3 (1902), 1–11.
Wang Ling, "On the Invention, and Use of Gunpowder and Firearms in China," *Isis*, 37, Nos. 109–10 (Pts. 3–4) (July 1947), 160–78.

Other sources on the use of rockets by the Chinese include eighteenth- and early nineteenth-century reports by French Jesuit missionaries and travelers. Among the more notable are:

Amiot, Joseph-Maria, *Art Militaire des Chinois* . . . Paris: Didot, 1772.

Gaubil, Antoine, *Histoire de Gentchiscan et de Toute la Dynastie de Mongous, ses Successeurs, Conquérants de la Chine.* Paris: Nyon, 1739.

Mailla, Joseph Anne-Marie Moriac de, *Histoire Génerâle de la Chine.* Paris: Pierres, 1777.

Ohsson, Constantine Mouradgea d', *Histoire des Mongols* . . . The Hague: Van Cleef, 1834.

Pauthier, Jean Pierre Guillaume, *Chine, ou Description Historique* . . . Paris: Firmin-Didot, 1821.

Joseph Needham, in his monumental multivolume *Science and Civilization in China* (Cambridge University Press, 1954–) gives some mention to gunpowder rockets in completed parts of the series, but, they are to be covered in full in Section 30, which will deal with military technology. Professor Needham wrote the authors that the subsection on gunpowder, including rockets, was begun in collaboration with Wang Ching-Ning, but is now being pursued by Ho Ping-Yö, Dean of the Faculty of Oriental Studies at Griffith University, Brisbane, Australia.

An important scholarly source on Chinese rocketry as well as Arab and Persian developments is:

Reinaud, Joseph Toussaint, and Ildephonse Favé, "Du Feu Grégeois, des Feux de Guerre, et des Origines de la Poudre à Canon Chez les Arabes, les Persans, et les Chinois," *Journal Asiatique,* 14, No. 10 (Oct. 1849), 257–327.

Arab developments alone are found in:

Reinaud, Joseph Toussaint, "De l'Art Militaire Chez les Arabes au Moyen Age," *Journal Asiatique,* 12, No. 9 (Sept. 1848), 193–237.

Two works are particularly valuable sources on Indian developments:

Gode, P. K., "The History of Fireworks in India Between A.D. 1400 and 1900," *Transaction,* No. 17 (Indian Institute of Culture) May 1953.

Katre, S. M., and P. K. Gode, *Use of Guns and Gunpowder in India from A.D. 1400 Onwards.* Bombay: Karnatak, 1939. (Indian and Iranian studies presented to Sir E. Denison Ross on his sixty-eighth birthday.)

The operational use of Indian rockets in the late eighteenth century is covered in many sources, a small selection of which follows:

Beatson, Alexander, *A View of the Origin and Conduct of the War with Tippoo Sultaun* . . . London: Nicol, 1900.

Bowring, Lewin B., *Haidar Ali and Tipu Sultan and the Struggle with the Musalman Powers of the South.* Oxford: Clarendon Press, 1899.

Diron, Alexander, *A Narrative of the Campaign in India, which Terminated the War with Tipoo Sultan.* London: Bulmer, 1793.

Hammick, Murray, ed., *Historical Sketches of the South of India, in an Attempt to Trace the History of Mysoor* . . . Mysore: Government Branch Press, 1930.

Hook, Theodore E., *The Life of General the Right Honourable Sir David Baird.* 2 vols. London: Bentley, 1832.

Munro, Innes, *A Narrative of the Military Operations on the Coromandel Coast* . . . London: Bentley, 1789.

Earlier Indian rockets are mentioned in such works as:

Egerton of Tatton, Lord, *A Description of Indian and Oriental Armour* . . . London: Allen, 1896.

Lane-Poole, Stanley, *Medieval India Under Mohammedan Rule, 712–1764.* New York: Putnam, 1903.

Oppert, Gustav, *On the Weapons, Army Organization, and Political Maxims of the Ancient Hindus.* London: Truebner, 1880.

Some of the more comprehensive works associated with Chapter 7, "Pioneering Modern Rocketry," follow. The historical books by Ley and by von Braun and Ordway already cited in this bibliography should also be consulted.

Ananoff, Alexandre, *L'Astronautique.* Paris: Librairie Arthème Fayard, 1950.

Baxter, James Phinney, 3rd, *Scientists Against Time.* Cambridge, Mass.: M.I.T. Press, 1946. (See chapter XIII, "Rockets.")

Benecke, Th., and A. W. Quirk, eds., *History of German Guided Missile Development.* Brunswick: Verlag E. Appelhaus, 1957.

Dornberger, Walter, *V-2.* New York: Viking, 1952.

Esnault-Pelterie, Robert, *L'Astronautique*. Paris: Lahure, 1930.

———, *L'Astronautique–Complément*. Paris: Société des Ingénieurs Civils de France, 1935.

———, *L'Exploration par Fusées de la Très Haute Atmosphère et la Possibilité des Voyages Interplanétaires*. Paris: Société Astronomique de France, 1927.

Gartmann, Heinz, *The Men Behind the Space Rockets*. New York: David McKay, 1956.

Goddard, Esther C., and G. Edward Pendray, eds., *The Papers of Robert H. Goddard*. 3 vols. New York: McGraw-Hill, 1968.

Goddard, Robert H., *A Method of Reaching Extreme Altitudes*. Washington, D.C.: Smithsonian Institution, 1919.

———, *Liquid-Propellant Rocket Development*. Washington, D.C.: Smithsonian Institution, 1936.

———, *Rocket Development*. Englewood Cliffs, N.J.: Prentice-Hall, 1948.

Hohmann, Walter, *Die Erreichbarkeit der Himmelskörper*. Munich: Oldenbourg, 1925.

Irving, David, *The Mare's Nest*. London: William Kimber, 1964.

Joubert de la Ferté, Sir Philip B., *Rocket*. New York: Philosophical Library, 1957.

Klee, Ernst, and Otto Merk, *The Birth of the Missile: The Secrets of Peenemünde*. New York: Dutton, 1965.

Kosmodemyansky, A., *Konstantin Tsiolkovsky*. Moscow: Foreign Languages Publishing House, 1956.

Lehman, Milton, *This High Man: The Life of Robert H. Goddard*. New York: Farrar, Straus, 1963.

Ley, Willy, *Die Fahrt ins Weltall*. Leipzig: Hachmeister und Thal, 1926.

———, ed., *Die Möglichkeit der Weltraumfahrt*. Leipzig: Hachmeister und Thal, 1928.

Lusar, Rudolf, *German Secret Weapons of the Second World War*. New York: Philosophical Library, 1959.

McGovern, James, *Crossbow and Overcast*. New York: William Morrow, 1964.

Nebel, Rudolf, *Die Narren von Tegel*. Düsseldorf: Droste, 1972.

Noordung, Hermann, *Das Problem der Befahrung des Weltraums*. Berlin: Richard Carl Schmidt, 1929.

Oberth, Hermann, *Die Rakete zu den Planetenräumen*. Munich: Oldenbourg, 1923.

———, *Wege zur Raumschiffahrt*. Munich: Oldenbourg, 1929.

Ohart, Theodore C., *Elements of Ammunition*. New York: Wiley, 1946.

Ordway, Frederick I., III, and Ronald C. Wakeford, *International Missile and Spacecraft Guide*. New York: McGraw-Hill, 1960.

Pendray, G. Edward, *The Coming Age of Rocket Power*. New York: Harper, 1945.

Rynin, N. A., *Interplanetary Flight and Communication*. 9 vols. Jerusalem: Israel Program for Scientific Translation, 1970–71. (Translated from Russian original published in Leningrad 1928–1932.)

Sänger, Eugen, *Raketenflugtechnik*. Munich: Oldenbourg, 1933.

———, and Irene Bredt, trans., *Rocket Drive for Long Range Bombers*. Whittier, Calif.: Robert Cornog, 1952.

Scherschevsky, Alexander Boris, *Die Rakete für Fahrt und Flug*. Berlin: Volckmann, 1929.

Tsiolkovsky, Konstantin Eduardovich, *Collected Works*. Washington, D.C.: NASA, 1965. Translation of *Sobranie Sochinenie*, published in Moscow in 1951, 1954 and 1959.

———, *Works on Rocket Technology*. Washington, D.C.: NASA, 1965.

Valier, Max, *Der Vorstoss in den Weltenraum*. Munich: Oldenbourg, 1928.

———, *Rakentenfahrt*. Munich: Oldenbourg, 1930.

Williams, Beryl, and Samuel Epstein, *Rocket Pioneers on the Road to Space*. New York: Julian Messner, 1955.

The bibliography for Chapters 8 and 9, "The Military Balance" and "The Quest for New Worlds," is combined, since post-World War II military missilery and space flight technology grew on the same base and share much of the same hardware. Again, the Ley and von Braun-Ordway histories provide comprehensive surveys, backed up by such sources as the regularly appearing volumes *Space Research*, published by the Committee on Space Research (better known as COSPAR), *Astronautical Research*, and earlier *Proceedings* of the annual meetings of the International Astronautical Federation, and the series *Advances in Space Science and Technology*, edited by Frederick I. Ordway III (Academic Press).

Abel, Elie, *The Missiles of October: The Cuban Missile Crisis 1962*. London: MacGibbon & Keye, 1966.

Adams, Carsbie C., Frederick I. Ordway III, Heyward E. Canney, and Ronald C. Wakeford, *Space Flight*. New York: McGraw-Hill, 1958.

Armstrong, Neil, Michael Collins, and Edwin E. Aldrin, Jr., *First on the Moon.* Boston: Little, Brown, 1970.

Baar, James, and W. E. Howard, *Combat Missilemen.* New York: Harcourt, Brace & World, 1961.

Baumgartner, John Standly, *The Lonely Warriors.* Los Angeles: Nash, 1970.

Bedini, Silvio A., Wernher von Braun, and Fred L. Whipple, *Moon: Man's Greatest Adventure.* New York: Abrams, 1971.

Bonestell, Chesley, and Arthur C. Clarke, *Beyond Jupiter: The Worlds of Tomorrow.* Boston: Little, Brown, 1973.

Boyd, R. L. F., and M. J. Seaton, eds., *Rocket Exploration of the Upper Atmosphere.* New York: Pergamon, 1954.

Bradbury, Ray, and others, *Mars and the Mind of Man.* New York: Harper & Row, 1973.

Braun, Wernher von, *First Men to the Moon.* New York: Holt, Rinehart and Winston, 1958.

———, *Mars Project.* Urbana: University of Illinois Press, 1953.

———, *Space Frontier.* New York: Holt, Rinehart and Winston, 1967, 1971.

Bush, Vannevar, *Modern Arms and Free Men.* New York: Simon & Schuster, 1949.

Clarke, Arthur C., *Interplanetary Flight.* New York: Harper, 1950.

———, *Man and Space.* New York: Time, Inc., 1964.

———, *The Promise of Space.* New York: Harper, 1968.

Cooper, Henry S. F., Jr., *Thirteen: The Flight That Failed.* New York: Dial, 1973.

Corliss, William R., *Scientific Satellites.* Washington, D.C.: NASA, 1967.

———, *Space Probes and Planetary Exploration.* Princeton, N.J.: Van Nostrand, 1965.

Emme, Eugene M., ed., *History of Rocket Technology.* Detroit: Wayne State University Press, 1964.

Gagarin, Yuri, *Road to the Stars.* Moscow: Foreign Languages Publishing House, 1962.

Gantz, K. F., *United States Air Force Report on the Ballistic Missile.* Garden City, N.Y.: Doubleday, 1958.

Gatland, Kenneth W., *Development of the Guided Missile.* New York: Philosophical Library, 1952.

Gavin, James M., *War and Peace in the Space Age.* New York: Harper, 1958.

Glasstone, Samuel, *Sourcebook of the Space Sciences.* Princeton, N.J.: Van Nostrand, 1965.

Glushko, V. P., *Development of Rocketry and Space Technology in the U.S.S.R.* Moscow: Novosti Press Publishing House, 1973.

Green, Constance McLaughlin, and Milton Lomask, *Vanguard: A History.* Washington, D.C.: Smithsonian Institution, 1971.

Hadley, Arthur T., *The Nation's Safety and Arms Controls.* New York: Viking, 1961.

Hirsch, Richard, and Joseph John Trento, *The National Aeronautics and Space Administration.* New York: Praeger, 1973.

Holder, William G., *Skylab, Pioneer Space Station.* Chicago: Rand McNally, 1974.

James, Peter N., *Soviet Conquest from Space.* New Rochelle, N.Y.: Arlington House, 1974.

Klass, Philip J., *Secret Sentries in Space.* New York: Random House, 1971.

Koelle, H. H., ed., *Handbook of Astronautical Engineering.* New York: McGraw-Hill, 1961.

Krieger, F. J., *Behind the Sputniks.* Washington, D.C.: Public Affairs Press, 1958.

Lasby, Clarence G., *Project Paperclip.* New York: Atheneum, 1971.

Ley, Willy, and Chesley Bonestell, *The Conquest of Space.* New York: Viking, 1949.

Ley, Willy, Wernher von Braun, and Chesley Bonestell, *The Exploration of Mars.* New York: Viking, 1960.

Logsdon, John M., *The Decision to Go to the Moon.* Cambridge, Mass.: M.I.T. Press, 1974.

Loosbrock, J. F., and others, eds., *Space Weapons—A Handbook of Military Astronautics.* New York: Praeger, 1959.

Medaris, John B., *Countdown for Decision.* New York: Putnam, 1960.

Newell, Homer E., Jr., *Sounding Rockets.* New York: McGraw-Hill, 1959.

Ordway, Frederick I., III, *Pictorial Guide to Planet Earth.* New York: Crowell, 1975.

Ordway, Frederick I., III, Carsbie C. Adams, and Mitchell R. Sharpe, *Dividends from Space.* New York: Crowell, 1971.

Ordway, Frederick I., III, James Patrick Gardner, and Mitchell R. Sharpe, *Basic Astronautics.* Englewood Cliffs, N.J.: Prentice-Hall, 1962.

Ordway, Frederick I., III, James Patrick Gardner, Mitchell R. Sharpe, and Ronald C. Wakeford, *Applied Astronautics*. Englewood Cliffs, N.J.: Prentice-Hall, 1963.

Ordway, Frederick I., III, and Ronald C. Wakeford, *International Missile and Spacecraft Guide*. New York: McGraw-Hill, 1960.

Parry, Albert, *Russia's Rockets and Missiles*. Garden City, N.Y.: Doubleday, 1960.

Petrov, G. I., ed., *Conquest of Outer Space in the USSR*. New Delhi: Amerind, 1973.

Riabchikov, Evgeny, *Russians in Space*. Garden City, N.Y.: Doubleday, 1971.

Rose, Ulrich Detlev, *Die Unheimlichen Waffen*. Munich: Schild Verlag, 1957.

Rosen, Milton, *The Viking Rocket Story*. New York: Harper, 1955.

Ruzic, Neil P., *Where the Winds Sleep*. Garden City, N.Y.: Doubleday, 1970.

Ryan, Cornelius, ed., *Across the Space Frontier*. New York: Viking, 1952.

———, *Conquest of the Moon*. New York: Viking, 1953.

Salkeld, Robert, *War and Space*. Englewood Cliffs, N.J.: Prentice-Hall, 1970.

Sänger, Eugen, *Space Flight: Countdown for the Future*. New York: McGraw-Hill, 1965.

Schwiebert, Ernest G., *A History of the U. S. Air Force Ballistic Missiles*. New York: Praeger, 1965.

Sharpe, Mitchell R., *Satellites and Probes*. London: Aldus, 1970.

———, *Living in Space: The Environment of the Astronaut*. Garden City, N.Y.: Doubleday, 1969.

———, *Yuri Gagarin, First Man in Space*. Huntsville, Ala.: Strode, 1969.

Stehling, Kurt R., *Project Vanguard*. Garden City, N.Y.: Doubleday, 1961.

Stuhlinger, Ernst, Frederick I. Ordway III, Jerry C. McCall, and George C. Bucher, eds., *Astronautical Engineering and Science*. New York: McGraw-Hill, 1963.

Swenson, Loyd S., Jr., James M. Grimwood, and Charles C. Alexander, *This New Ocean: A History of Project Mercury*. Washington, D.C.: NASA, 1966.

Thomas, Shirley, *Men of Space*. Vols. 1–8. Philadelphia: Chilton, 1960–68.

Titov, Gherman, and Martin Caiden, *I Am Eagle*. Indianapolis: Bobbs-Merrill, 1962.

Vaeth, J. Gordon, *200 Miles Up*. New York: Ronald, 1955.

Van Allen, James A., ed., *Scientific Uses of Earth Satellites.* Ann Arbor: University of Michigan Press, 1956.

Watson, Clement Hayes, ed., *Adventures in Partnership.* Danbury, Conn.: Danbury Printing and Litho Co., 1972.

Young, Dayton, *One More Chance.* New York: Exposition Press, 1973.

Young, Hugo, Bryan Silcock, and Peter Dunn, *Journey to Tranquillity.* Garden City, N.Y.: Doubleday, 1970.

Zaehringer, Alfred J., *Soviet Space Technology.* New York: Harper & Row, 1961.

INDEX

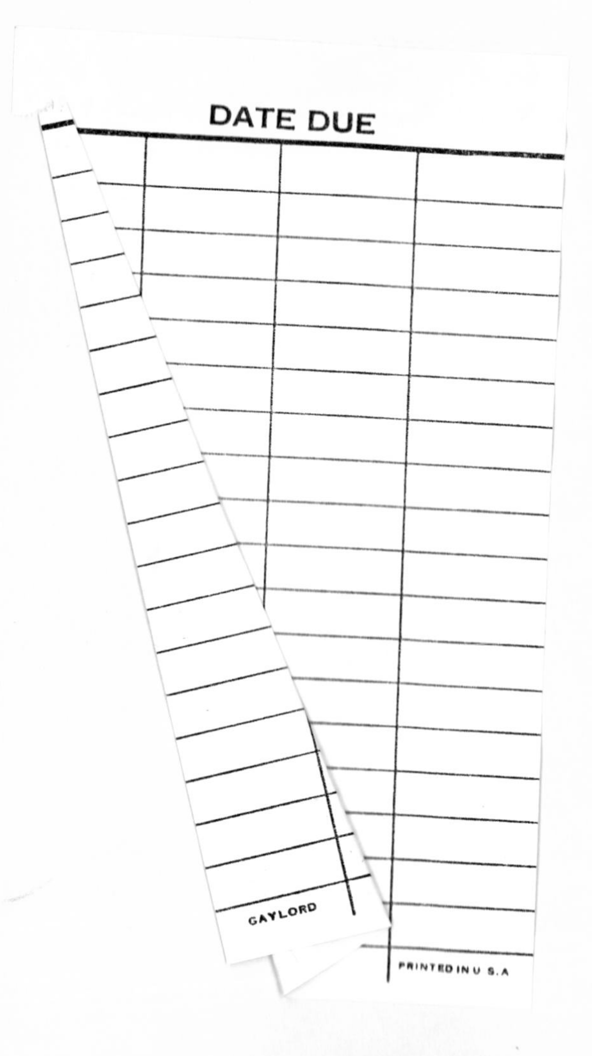
DATE DUE
GAYLORD
PRINTED IN U.S.A.